MAKING SENSE OF DATA II

MAKING SENSE OF DATA II

A Practical Guide to Data Visualization, Advanced Data Mining Methods, and Applications

GLENN J. MYATT

WAYNE P. JOHNSON

WILEY

A JOHN WILEY & SONS, INC., PUBLICATION

Published by John Wiley & Sons, Inc., Hoboken, New Jersey

Published simultaneously in Canada

For general information on our other products and services or for technical support, please contact our Customer Care Department within the United States at (800) 762-2974, outside the United States at (317) 572-3993 or fax (317) 572-4002.

Wiley also publishes its books in variety of electronic formats. Some content that appears in print may not be available in electronic formats. For more information about Wiley products, visit our web site at www.wiley.com.

Library of Congress Cataloging-in-Publication Data:

Myatt, Glenn J., 1969-

   Making sense of data II: a practical guide to data visualization, advanced data mining methods, and applications/Glenn J. Myatt, Wayne P. Johnson.

        p. cm.

   Making sense of data 2

   Includes bibliographical references and index.

   ISBN 978-0-470-22280-5 (pbk.)

1. Data mining. 2. Information visualization. I. Johnson, Wayne P. II. Title. III. Title: Making sense of data 2.

   QA76.9.D343M93 2008

   005.74--dc22

                                                            2008024103

10   9   8   7   6   5   4   3   2

CONTENTS

PREFACE

The purpose of this book is to outline a diverse range of commonly used approaches to making and communicating decisions from data, using data visualization, clustering, and predictive analytics. The book relates these topics to how they can be used in practice in a variety of ways. First, the methods outlined in the book are discussed within the context of a data mining process that starts with defining the problem and ends with deployment of the results. Second, each method is outlined in detail, including a discussion of when and how they should be used. Third, examples are provided throughout to further illustrate how the methods operate. Fourth, there is a detailed discussion of applications in which these approaches are being applied today. Finally, software called Traceis™, which can be used with the examples in the book or with data sets of interest to the reader, is available for downloading from a companion website.

The book is aimed towards professionals in any discipline who are interested in making decisions from data in addition to understanding how data mining can be used. Undergraduate and graduate students taking courses in data mining through a Bachelors, Masters, or MBA program could use the book as a resource. The approaches have been outlined to an extent that software professionals could use the book to gain insight into the principles of data visualization and advanced data mining algorithms in order to help in the development of new software products.

The book is organized into five chapters and two appendices.

- *Chapter 1—Introduction*: The first chapter reviews the material in the book within the context of the overall data mining process. Defining the problem, preparing the data, performing the analysis, and deploying any results are critical steps. When and how each of the methods described in the book can be applied to this process are described.

- *Chapter 2—Data Visualization*: The second chapter reviews principles and methods for understanding and communicating data through the use of data visualizations. The chapter outlines ways of visualizing single variables, the relationships between two or more variables, groupings in the data, along with dynamic approaches to interacting with the data through graphical user interfaces.

- *Chapter 3—Clustering*: Chapter 3 outlines in detail common approaches to clustering data sets and includes a detailed explanation of methods for determining the distance between observations and techniques for clustering observations. Three popular clustering approaches are discussed: agglomerative hierarchical clustering, partitioned-based clustering, and fuzzy clustering.

- *Chapter 4—Predictive Analytics*: The ability to calculate estimates and forecasts or assign observations to specific classes using models is discussed. The chapter discusses how to build and assess models, along with a series of methods that can be used in a variety of situations to build models: multiple linear regression, discriminant analysis, logistic regression, and naive Bayes.

- *Chapter 5—Applications*: This chapter provides a snapshot of some of the current uses of data mining in a variety of industries. It also offers an overview of how data mining can be applied to topics where the primary focus is not tables of data, such as the processing of text documents and chemicals. A number of case studies illustrating the use of data mining are outlined.

- *Appendix A—Matrices*: This section provides an overview of matrices to use in connection with Chapters 3 and 4.

- *Appendix B—Software*: This appendix provides a detailed explanation of the capabilities of the Traceis software, along with a discussion of how to access, run, and use the software.

It is assumed that the reader of the book has a basic understanding of the principles of data mining. An overview has been given in a previously published book called *Making Sense of Data: A Practical Guide to Exploratory Data Analysis and Data Mining*, which outlines a simple process along with a core set of data analysis and data mining methods to use, explores additional and more advanced data mining methods, and describes the application of data mining in different areas.

Data mining issues and approaches from a number of perspectives are discussed in this book. The visualization and exploration of data is an essential component and the principles of graphics design and visualization of data are outlined to most effectively see and communicate the contents of the data. The methods outlined in Chapters 3 and 4 are described in such a way as to be used immediately in connection with any problem. The software provides a complementary tool, since one of the best ways to understand how these methods works is to use them on data, especially your own data. The Further Readings section of each chapter suggests material for further reading on topics related to the chapter.

Companion Website

Accompanying this book is a website:

http://www.makingsenseofdata.com/

containing additional resources to help in understanding how to implement the topics covered in this book. Included on the website is a software download for the Traceis software.

In putting this book together, we would like to thank the following individuals for their considerable help: Dr. Paul Blower, Dr. Satish Nargundkar, Kristen Blankley, and Vinod Chandnani. We would also like to thank all those involved in the review process for the book. Finally, we would like to thank the staff at John Wiley & Sons, particularly Susanne Steitz-Filler, for all their help and support throughout the entire project.

GLENN J. MYATT
WAYNE P. JOHNSON

Jasper, Georgia
November 2008

CHAPTER 1

INTRODUCTION

1.1 OVERVIEW

A growing number of fields, in particular the fields of business and science, are turning to data mining to make sense of large volumes of data. Financial institutions, manufacturing companies, and government agencies are just a few of the types of organizations using data mining. Data mining is also being used to address a wide range of problems, such as managing financial portfolios, optimizing marketing campaigns, and identifying insurance fraud. The adoption of data mining techniques is driven by a combination of competitive pressure, the availability of large amounts of data, and ever increasing computing power. Organizations that apply it to critical operations achieve significant returns. The use of a process helps ensure that the results from data mining projects translate into actionable and profitable business decisions. The following chapter summarizes four steps necessary to complete a data mining project: (1) definition, (2) preparation, (3) analysis, and (4) deployment. The methods discussed in this book are reviewed within this context. This chapter concludes with an outline of the book's content and suggestions for further reading.

1.2 DEFINITION

The first step in any data mining process is to define and plan the project. The following summarizes issues to consider when defining a project:

- *Objectives*: Articulating the overriding business or scientific objective of the data mining project is an important first step. Based on this objective, it is also important to specify the success criteria to be measured upon delivery. The project should be divided into a series of goals that can be achieved using available data or data acquired from other sources. These objectives and goals should be understood by everyone working on the project or having an interest in the project's results.

- *Deliverables*: Specifying exactly what is going to be delivered sets the correct expectation for the project. Examples of deliverables include a report outlining the results of the analysis or a predictive model (a mathematical model that estimates critical data) integrated within an operational system. Deliverables also

Making Sense of Data II. By Glenn J. Myatt and Wayne P. Johnson identify who will use the results of the analysis and how they will be delivered. Consider criteria such as the accuracy of the predictive model, the time required to compute, or whether the predictions must be explained.

- *Roles and Responsibilities*: Most data mining projects involve a cross-disciplinary team that includes (1) experts in data analysis and data mining, (2) experts in the subject matter, (3) information technology professionals, and (4) representatives from the community who will make use of the analysis. Including interested parties will help overcome any potential difficulties associated with user acceptance or deployment.

- *Project Plan*: An assessment should be made of the current situation, including the source and quality of the data, any other assumptions relating to the data (such as licensing restrictions or a need to protect the confidentiality of the data), any constraints connected to the project (such as software, hardware, or budget limitations), or any other issues that may be important to the final deliverables. A timetable of events should be implemented, including the different stages of the project, along with deliverables at each stage. The plan should allot time for cross-team education and progress reviews. Contingencies should be built into the plan in case unexpected events arise. The timetable can be used to generate a budget for the project. This budget, in conjunction with any anticipated financial benefits, can form the basis for a cost–benefit analysis.

1.3   PREPARATION

1.3.1   Overview

Preparing the data for a data mining exercise can be one of the most time-consuming activities; however, it is critical to the project's success. The quality of the data accumulated and prepared will be the single most influential factor in determining the quality of the analysis results. In addition, understanding the contents of the data set in detail will be invaluable when it comes to mining the data. The following section outlines issues to consider when accessing and preparing a data set. The format of different sources is reviewed and includes data tables and nontabular information (such as text documents). Methods to categorize and describe any variables are outlined, including a discussion regarding the scale the data is measured on. A variety of descriptive statistics are discussed for use in understanding the data. Approaches to handling inconsistent or problematic data values are reviewed. As part of the preparation of the data, methods to reduce the number of variables in the data set should be considered, along with methods for transforming the data that match the problem more closely or to use with the analysis methods. These methods are reviewed. Finally, only a sample of the data set may be required for the analysis, and techniques for segmenting the data are outlined.

1.3.2 Accessing Tabular Data

Tabular information is often used directly in the data mining project. This data can be taken directly from an operational database system, such as an ERP (enterprise resource planning) system, a CRM (customer relationship management) system, SCM (supply chain management) system, or databases containing various trans- actions. Other common sources of data include surveys, results from experiments, or data collected directly from devices. Where internal data is not sufficient for the objective of the data mining exercise, data from other sources may need to be acquired and carefully integrated with existing data. In all of these situations, the data would be formatted as a table of observations with information on different variables of interest. If not, the data should be processed into a tabular format.

Preparing the data may include joining separate relational tables, or concatenat- ing data sources; for example, combining tables that cover different periods in time. In addition, each row in the table should relate to the entity of the project, such as a customer. Where multiple rows relate to this entity of interest, generating a summary table may help in the data mining exercise. Generating this table may involve calcu- lating summarized data from the original data, using computations such as sum, mode (most common value), average, or counts (number of observations). For example, a table may comprise individual customer transactions, yet the focus of the data mining exercise is the customer, as opposed to the individual transactions. Each row in the table should refer to a customer, and additional columns should be gener- ated by summarizing the rows from the original table, such as total sales per product. This summary table will now replace the original table in the data mining exercise.

Many organizations have invested heavily in creating a high-quality, consoli- dated repository of information necessary for supporting decision-making. These repositories make use of data from operational systems or other sources. Data ware- houses are an example of an integrated and central corporate-wide repository of decision-support information that is regularly updated. Data marts are generally smal- ler in scope than data warehouses and usually contain information related to a single business unit. An important accompanying component is a metadata repository, which contains information about the data. Examples of metadata include where the data came from and what units of measurements were used.

1.3.3 Accessing Unstructured Data

In many situations, the data to be used in the data mining project may not be represented as a table. For example, the data to analyze may be a collection of documents or a sequence of page clicks on a particular web site. Converting this type of data into a tabular format will be necessary in order to utilize many of the data mining approaches described later in this book. Chapter 5 describes the use of nontabular data in more detail.

1.3.4 Understanding the Variables and Observations

Once the project has been defined and the data acquired, the first step is usually to understand the content in more detail. Consulting with experts who have knowledge about how the data was collected as well as the meaning of the data is invaluable. Certain assumptions may have been built into the data, for example specific values may have particular meanings. Or certain variables may have been derived from others, and it will be important to understand how they were derived. Having a thorough understanding of the subject matter pertaining to the data set helps to explain why specific relationships are present and what these relationships mean.

(As an aside, throughout this book variables are presented in italics.)

An important initial categorization of the variables is the scale on which they are measured. *Nominal* and *ordinal* scales refer to variables that are *categorical*, that is, they have a limited number of possible values. The difference is that ordinal variables are ordered. The variable *color* which could take values black, white, red, and so on, would be an example of a nominal variable. The variable *sales*, whose values are low, medium, and high, would be an example of an ordinal scale, since there is an order to the values. *Interval* and *ratio* scales refer to variables that can take any *continuous* numeric value; however, ratio scales have a natural zero value, allowing for a calculation of a ratio. Temperature measured in Fahrenheit or Celsius is an example of an interval scale, as it can take any continuous value within a range. Since a zero value does not represent the absence of temperature, it is classified as an interval scale. However, temperatures measured in degrees Kelvin would be an example of a ratio scale, since zero is the lowest temperature. In addition, a bank balance would be an example of a ratio scale, since zero means no value.

In addition to describing the scale on which the individual variables were measured, it is also important to understand the frequency distribution of the variable (in the case of interval or ratio scaled variables) or the various categories that a nominal or ordinal scaled variable may take. Variables are usually examined to understand the following:

- *Central Tendency*: A number of measures for the central tendency of a variable can be calculated, including the *mean* or average value, the *median* or the middle number based on an ordering of all values, and the *mode* or the most common value. Since the mean is sensitive to outliers, the *trimmed mean* may be considered which refers to a mean calculated after excluding extreme values. In addition, median values are often used to best represent a central value in situations involving outliers or skewed data.

- *Variation*: Different numbers show the variation of the data set's distribution. The minimum and maximum values describe the entire range of the variable. Calculating the values for the different quartiles is helpful, and the calculation determines the points at which 25% (Q1), 50% (Q2), and 75% (Q3) are found in the ordered values. The *variance* and *standard deviation* are usually calculated to quantify the data distribution. Assuming a normal distribution, in the case of standard deviation, approximately 68% of all observations fall within one standard deviation of the mean, and approximately 95% of all observations fall within two standard deviations of the mean.

- *Shape*: There are a number of metrics that define the shape and symmetry of the frequency distribution, including *skewness*, a measure of whether a variable is skewed to the left or right, and *kurtosis*, a measure of whether a variable has a flat or pointed central peak.

Graphs help to visualize the central tendency, the distribution, and the shape of the frequency distribution, as well as to identify any outliers. A number of graphs that are useful in summarizing variables include: frequency histograms, bar charts, frequency polygrams, and box plots. These visualizations are covered in detail in the section on univariate visualizations in Chapter 2.

Figure 1.1 illustrates a series of statistics calculated for a particular variable (*percentage body fat*). In this example, the variable contains 251 observations, and the most commonly occurring value is 20.4 (mode), the median is 19.2, and the average or mean value is 19.1. The variable ranges from 0 to 47.5, with the point at which 25% of the ordered values occurring at 12.4, 50% at 19.2 (or median), and 75% at 25.3. The variance is calculated to be 69.6, and the standard deviation at 8.34, that is, approximately 68% of observations occur ± 8.34 from the mean (10.76–28.44), and approximately 95% of observations occur ± 16.68 from the mean (2.42–35.78).

At this point it is worthwhile taking a digression to explain terms used for the different roles variables play in building a prediction model. The *response* variable, also referred to as the *dependent* variable, the *outcome*, or *y-variable*, is the variable any model will attempt to predict. *Independent* variables, also referred to as *descriptors, predictors,* or *x-variables*, are the fields that will be used in building the model. *Labels*, also referred to as *record identification*, or *primary key*, is a unique value corresponding to each individual row in the table. Other variables may be present in the table that will not be used in any model, but which can still be used in explanations.

During this stage it is also helpful to begin exploring the data to better understand its features. Summary tables, matrices of different graphs, along with interactive techniques such as brushing, are critical data exploration tools. These tools are described in Chapter 2 on data visualization. Grouping the data is also helpful to understand the general categories of observations present in the set. The visualization of groups is presented in Chapter 2, and an in-depth discussion of clustering and grouping methods is provided in Chapter 3.

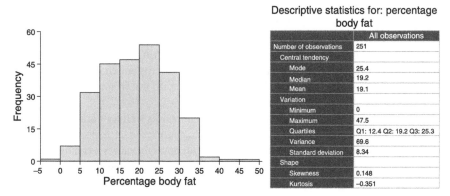

Figure 1.1   Descriptive statistics and a histogram

1.3.5 Data Cleaning

Having extracted a table containing observations (represented as rows) and variables (represented as columns), the next step is to clean the data table, which often takes a considerable amount of time. Some common cleaning operations include identifying (1) errors, (2) entries with no data, and (3) entries with missing data. Errors and missing values may be attributable to the original collection, the transmission of the information, or the result of the preparation process.

Values are often missing from the data table, but a data mining approach cannot proceed until this issue is resolved. There are five options: (1) remove the entire observation from the data table; (2) remove the variable (containing the missing values) from the data table; (3) replace the missing value manually; (4) replace the value with a computed value, for example, the variable's mean or mode value; and (5) replace the entry with a predicted value based on a generated model using other fields in the data table. Different approaches for generating predictions are described in Chapter 4 on Predictive Analytics. The choice depends on the data set and the problem being addressed. For example, if most of the missing values are found in a single variable, then removing this variable may be a better option than removing the individual observations.

A similar situation to missing values occurs when a variable that is intended to be treated as a numeric variable contains text values, or specific numbers that have specific meanings. Again, the five choices previously outlined above may be used; however, the text or the specific number value may suggest numeric values to replace them with. Another example is a numeric variable where values below a threshold value are assigned a text string such as "$<10^{**}-9$." A solution for this case might be to replace the string with the number 0.000000001.

Another problem occurs when values within the data tables are incorrect. The value may be problematic as a result of an equipment malfunction or a data entry error. There are a number of ways to help identify errors in the data. Outliers in the data may be errors and can be found using a variety of methods based on the variable, for example, calculating a *z-score* for each value that represents the number of standard deviations the value is away from the mean. Values greater than plus or minus three may be considered outliers. In addition, plotting the data using a box plot or a frequency histogram can often identify data values that significantly deviate from the mean. For variables that are particularly noisy, that is they contain some degree of errors, replacing the variable with a binned version that more accurately represents the variation of the data may be necessary. This process is called *data smoothing*. Other methods, such as data visualization, clustering, and regression models (described in Chapters 2–4) can also be useful to identify anomalous observations that do not look similar to other observations or that do not fit a trend observed for the majority of the variable's observations.

Looking for values that deviate from the mean works well for numeric variables; however, a different strategy is required to handle categorical data, especially where all data values are nonnumeric. Looking at the list of all possible values a variable can take helps to eliminate and/or consolidate values where more than one value has the same meaning, which might happen, for example, in a categorical variable. Even though a data value may look different from other values in the variable, the data may, in fact, be correct, so it is important to consult with an expert.

Problems can also arise when data from multiple sources is integrated and inconsistencies are introduced. Different sources may have values for the same variables; however, the values may have been recorded using different units of measurement and hence must be standardized to a single unit of measurement. Different sources of data may contain the same observation. Where the same observation has the same values for all variables, removing one of the observations is the most straightforward approach. Where the observations have different values, choosing which observation to keep is more challenging and best decided by someone who is able to assess the most trusted source. Other common problems when dealing with integrated data concern assessing how up-to-date the observations are and whether the quality is the same across different sources of data. Where observations are taken from different sources, retaining information on the source for future reference is prudent.

1.3.6 Transformation

In many situations, it is necessary to create new variables from existing columns of data to reflect more closely the purpose of the project or to enhance the quality of the predictions. For example, creating a new column *age* from an existing column *date of birth*, or computing an average from a series of experimental runs might be helpful. The data may also need to be transformed in order to be used with a particular analysis technique. There are six common transformations:

1. *Creating Dummy Variables*: A variable measured on a nominal or ordinal scale is usually converted into a series of *dummy* variables for use within data mining methods that require numbers. Each category is usually converted to a variable with one of two values: a one when the value is present in the observation and a zero when it is absent. Since this method would generate a new variable for each category, care should be taken when using all these columns with various methods, such as multiple linear regression or logistic regression (discussed in Chapter 4). These methods are sensitive to issues relating to *colinearity* (a high degree of correlation between variables), and hence including all variables would introduce a problem for these methods. When a final variable can be deduced from the other variables, there is no need to include the final variable. For example, the variable *color* whose values are black, white, and red could be translated into three dummy variables, one for each of the three values. Each observation would have a value one for the color corresponding to the row, and zero corresponding to the other two colors. Since the red column can be derived from the other two columns, only black and white columns are needed. The use of dummy variables is illustrated in the case studies in Chapter 5.

2. *Reducing the Number of Categories*: A categorical variable may be comprised of many different values, and using the variable directly may not draw any meaningful conclusions; however, generalizing the values may generate useful conclusions. This can be achieved through a manual definition of a

concept hierarchy or assisted using automated approaches. References in the further readings section of this chapter discuss this further, along with Appendix B (Software). For example, a variable comprising street names may be more valuable if it is generalized to the town containing those streets. This may be achieved through the construction of a concept hierarchy, where individual street names map on to the town names. In this case, there will be more observations for a particular town which hopefully result in more interesting conclusions.

3. *Create Bins for Continuous Variables*: To facilitate the use of a continuous variable within methods that require categorical variables (such as the association rules method), or to perform data smoothing, a continuous variable could be divided into a series of contiguous ranges or *bins*. Each of the observation's values would then be assigned to a specific bin, and potentially assigned a value such as the bin's mean. For example, a variable *temperature* with values ranging from 0 to 100, may be divided into a series of bins: $0-10$, $10-20$, and so on. A value could be assigned as each bin's mid-point. There are a variety of manual or automated approaches, and references to them are provided in the further readings section of this chapter, as well as in cases in Chapter 5 (Applications) and Appendix B (Software).

4. *Mapping the Data to a Normal Distribution*: Certain modeling approaches require that the frequency distribution of the variables approximate a normal distribution, or a bell-shaped curve. There are a number of common transformations that can be applied to a variable to achieve this. For example, a *Box-Cox* transformation or a *log* transformation may be used to generate a new variable where the data more closely follows the bell-shaped curve of a normal distribution. The Further Reading section, as well as Appendix B, provide more details related to this subject.

5. *Standardizing the Variables to a Consistent Range*: In order to treat different variables with the same weight, a scheme for normalizing the variables to the same range is often used, such as between zero and one. *Min−max*, *z-score*, and *decimal scaling* are examples of approaches to normalizing data to a specific, common range. As an example, a data set containing the variables *age* and *bank account balance* may be standardized using the *min−max* normalization to a consistent range of zero to one. These new variables make possible the consistent treatment of variables within methods, such as clustering, which utilizes distances between variables. If these two variables were not on a standard range, the *bank account balance* variable would, for the most part, be more influential than the *age* variable.

6. *Calculating Terms to Enhance Prediction*: To improve prediction, certain variables may be combined, or the variables may be transformed using some sort of mathematical operation. This may, for example, allow the more accurate modeling of nonlinear relationships. Some commonly used mathematical operations include square, cube, and square root. Appendix B and the Further Reading section of this chapter provide more details and references on this subject.

1.3.7 Variable Reduction

A data set with a large number of variables can present a number of issues within data mining techniques, including the problems of over fitting and model reliability, as well as potential computational problems. In this situation, selecting a subset of the variables will be important. This is sometimes referred to as *feature selection*. An expert with knowledge of the subject matter may be able to identify easily the variables that are not relevant to the problem. Variables that contain the same value for almost all observations do not provide much value and could be removed at this stage. In addition, categorical variables where the majority of observations have different values might not be useful within the analysis, but they may be useful to define the individual observations.

Understanding how the data will be used in a deployment scenario can also be useful in determining which variables to use. For example, the same independent variables must be gathered within a deployment scenario. However, it may be not practical to collect all the necessary data values, so it may be best to eliminate these variables at the beginning. For example, when developing a model to estimate hypertension propensity within a large patient population, a training set may include a variable *percentage body fat* as a relevant variable. The accurate measurement of this variable, however, is costly, and collecting it for the target patient population would be prohibitive. Surrogates, such as a skin-fold measurement, may be collected more easily and could be used instead of *percentage body fat*.

Additionally, examining the relationships between the variables is important. When building predictive models, there should be little relationship between the variables used to build the model. Strong relationships between the independent variables and the response variables are important and can be used to prioritize the independent variables. Bivariate data visualizations, such as scatterplot matrices, are important tools, and they are described in greater detail in Chapter 2. Calculating a correlation coefficient for each pair of continuous variables and presenting these calculations in a table can also be helpful in understanding the linear relationships between all pairs of variables, as shown in Fig. 1.2. For example, there is a strong negative linear relationship between *percentage body fat* and *density* (-0.988), a strong positive linear relationship between *abdomen (cm)* and *chest (cm)* (0.916), and a lack of a clear linear relationship between *height (inches)* and *percentage body fat* since it is close to zero.

	Density	Percentage body fat	Weight, lb	Height, in	Chest, cm	Abdomen, cm	Thigh, cm
Density	1	−0.988	−0.592	0.0375	−0.683	−0.798	−0.547
Percentage body fat	−0.988	1	0.611	−0.0234	0.703	0.813	0.554
Weight, lb	−0.592	0.611	1	0.489	0.894	0.888	0.87
Height, in	0.0375	−0.0234	0.489	1	0.228	0.192	0.344
Chest, cm	−0.683	0.703	0.894	0.228	1	0.916	0.732
Abdomen, cm	−0.798	0.813	0.888	0.192	0.916	1	0.766
Thigh, cm	−0.547	0.554	0.87	0.344	0.732	0.766	1

Figure 1.2 Matrix of correlation coefficients

Other techniques, such as *principal component analysis*, can also be used to reduce the number of continuous variables. The relationships between categorical independent variables can be assessed using statistical tests, such as the chi-square test. Decision trees are also useful for understanding important variables. Those chosen by the method that generates the tree are likely to be important variables to retain. Subsets of variables can also be assessed when optimizing the parameters to a data mining algorithm. For example, different combinations of independent variables can be used to build models, and those giving the best results should be retained. Methods for selecting variables are discussed in Chapter 4 on Predictive Analytics.

1.3.8   Segmentation

Using the entire data set is not always necessary, or even practical, especially when the number of observations is large. It may be possible to draw the same conclusions more quickly using a subset. There are a number of ways of selecting subsets. For example, using a random selection is often a good approach. Another method is to partition the data, using methods such as clustering, and then select an observation from each partition. This ensures the selection is representative of the entire collection of observations.

In situations where the objective of the project is to model a rare event, it is often useful to bias the selection of observations towards incorporating examples of this rare event in combination with random observations of the remaining collection. This method is called *balanced sampling*, where the response variable is used to drive how the partitioning of the data set takes place. For example, when building a model to predict insurance fraud, an initial training data set may only contain 0.1% fraudulent vs 99.9% nonfraudulent claims. Since the objective is the identification of fraudulent claims, a new training set may be constructed containing a better balance of fraudulent to nonfraudulent examples. This approach would result in improved models for identifying fraudulent claims; however, it may reduce the overall accuracy of the model. This is an acceptable compromise in this situation.

When samples are pulled from a larger set of data, comparing statistics of the sample to the original set is important. The minimum and maximum values, along with mean, median, and mode value, as well as variance and standard deviations, are a good start for comparing continuous variables. Statistical tests, such as the *t-test*, can also be used to assess the significance of any difference. When looking at categorical variables, the distribution across the different values should be similar. Generating a contingency table for the two sets can also provide insight into the distribution across different categories, and the chi-square test can be useful to quantify the differences.

Chapter 3 details methods for dividing a data set into groups, Chapter 5 discusses applications where this segmentation is needed, and Appendix B outlines software used to accomplish this.

1.3.9   Preparing Data to Apply

Having spent considerable effort preparing a data set ready to be modeled, it is also important to prepare the data set that will be scored by the prediction model in the same manner. The steps used to access, clean, and transform the training data should be repeated for those variables that will be applied to the model.

1.4   ANALYSIS

1.4.1   Data Mining Tasks

Once a data set is acquired and prepared for analysis, the next step is to select the methods to use for data mining. These methods should match the problem outlined earlier and the type of data available. The preceding exploratory data analysis will be especially useful in prioritizing different approaches, as information relating to data set size, level of noise, and a preliminary understanding of any patterns in the data can help to prioritize different approaches. Data mining tasks primarily fall into two categories:

- *Descriptive*: This refers to the ability to identify interesting facts, patterns, trends, relationships, or anomalies in the data. These findings should be nontrivial and novel, as well as valuable and actionable, that is, the information can be used directly in taking an action that makes a difference to the organization. Identifying patterns or rules associated with fraudulent insurance claims would be an example of a descriptive data mining task.
- *Predictive*: This refers to the development of a model of some phenomena that will enable the estimation of values or prediction of future events with confidence. For example, a prediction model could be generated to predict whether a cell phone subscriber is likely to change service providers in the near future. A predictive model is typically a mathematical equation that is able to calculate a value of interest (response) based on a series of independent variables.

Descriptive data mining usually involves grouping the data and making assessments of the groups in various ways. Some common descriptive data mining tasks are:

- *Associations*: Finding associations between multiple items of interest within a data set is used widely in a variety of situations, including data mining retail or marketing data. For example, online retailers determine product combinations purchased by the same set of customers. These associations are subsequently used when a shopper purchases specific products, and alternatives are then suggested (based on the identified associations). Techniques such as association rules or decision trees are useful in identifying associations within the data. These approaches are covered in Myatt (2007).
- *Segmentation*: Dividing a data set into multiple groups that share some common characteristic is useful in many situations, such as partitioning the market for a product based on customer profiles. These partitions help in developing targeted marketing campaigns directed towards these groups. Clustering methods are widely used to divide data sets into groups of related observations, and different approaches are described in Chapter 3.

- *Outliers*: In many situations, identifying unusual observations is the primary focus of the data mining exercise. For example, the problem may be defined as identifying fraudulent credit card activity; that is, transactions that do not follow an established pattern. Again, clustering methods may be employed to identify groups of observations; however, smaller groups would now be considered more interesting, since they are a reflection of unusual patterns of activity. Clustering methods are discussed in Chapter 3.

The two primary predictive tasks are:

- *Classification*: This is when a model is built to predict a categorical variable. For example, the model may predict whether a customer will or will not buy a particular product. Methods such as logistic regression, discriminant analysis, and naive Bayes classifiers are often used and these methods are outlined in Chapter 4 on Predictive Analytics.

- *Regression*: This is also referred to as *estimation*, *forecasting*, or *prediction*, and it refers to building models that generate an estimation or prediction for a continuous variable. A model that predicts the sales for a given quarter would be an example of a regression predictive task. Methods such as multiple linear regression are often used for this task and are discussed in Chapter 4.

1.4.2 Optimization

Any data mining analysis, whether it is finding patterns and trends or building a predictive model, will involve an iterative process of trial-and-error in order to find an optimal solution. This optimization process revolves around adjusting the following in a controlled manner:

- *Methods*: To accomplish a data mining task, many potential approaches may be applied; however, it is not necessarily known in advance which method will generate an optimal solution. It is therefore common to try different approaches and select the one that produces the best results according to the success criteria established at the start of the project.

- *Independent Variables*: Even though the list of possible independent variables may have been selected in the data preparation step, one way to optimize any data mining exercise is to use different combinations of independent variables. The simplest combinations of independent variables that produced the optimal predictive accuracy should be used in the final model.

- *Parameters*: Many data mining methods require parameters to be set that adjust exactly how the approach operates. Adjusting these parameters can often result in an improvement in the quality of the results.

1.4.3 Evaluation

In order to assess which data mining approach is the most promising, it is important to objectively and consistently assess the various options. Evaluating the different approaches also helps set expectations concerning possible performance levels during deployment. In evaluating a predictive model, different data sets should be used to build the model and to test the performance of the model, thus ensuring that the model has not overfitted the data set from which it is learning. Chapter 4 on Predictive Analytics outlines methods for assessing generated models. Assessment of the results from descriptive data mining approaches should reflect the objective of the data mining exercise.

1.4.4 Model Forensics

Spending time looking at a working model to understand when or why a model does or does not work is instructive, especially looking at the false positives and false negatives. Clustering, pulling out rules associated with these errors, and visualizing the data, may be useful in understanding when and why the model failed. Exploring this data may also help to understand whether additional data should be collected. Data visualizations and clustering approaches, described in Chapters 2 and 3, are useful tools to accomplish model forensics as well as to help communicate the results.

1.5 DEPLOYMENT

The discussion so far has focused on defining and planning the project, acquiring and preparing the data, and performing the analysis. The results from any analysis then need to be translated into tangible actions that impact the organization, as described at the start of the project. Any report resulting from the analysis should make its case and present the evidence clearly. Including the user of the report as an interested party to the analysis will help ensure that the results are readily understandable and usable by the final recipient.

One effective method of deploying the solution is to incorporate the analysis within existing systems, such as ERP or CRM systems, that are routinely used by the targeted end-users. Examples include using scores relating to products specific customers are likely to buy within a CRM system or using an insurance risk model within online insurance purchasing systems to provide instant insurance quotes. Integrating any externally developed models into the end-user system may require adoption of appropriate standards such as Object Linking and Embedding, Database for Data Mining (Data Mining OLE DB) which is an application programming interface for relational databases (described in Netz et al., 2001), Java Data Mining application programming interface standard (JSR-73 API; discussed in Hornick et al., 2006), and Predictive Model Markup Language (PMML; also reviewed in Hornick et al., 2006). In addition, the models may need to be integrated with current systems that are able to extract data from the current database and build the models automatically.

Other issues to consider when planning a deployment include:

- *Model Life Time*: A model may have a limited lifespan. For example, a model that predicts stock performance may only be useful for a limited time period, and it will need to be rebuilt regularly with current data in order to remain useful.

- *Privacy Issues*: The underlying data used to build models or identify trends may contain sensitive data, such as information identifying specific customers. These identities should not be made available to end users of the analysis, and only aggregated information should be provided.

- *Training*: Training end-users on how to interpret the results of any analysis may be important. The end-user may also require help in using the results in the most effective manner.

- *Measuring and Monitoring*: The models or analysis generated as a result of the project may have met specific evaluation metrics. When these models are deployed into practical situations, the results may be different for other unanticipated reasons. Measuring the success of the project in the field may expose an issue unrelated to the model performance that impacts the deployed results.

1.6   OUTLINE OF BOOK

1.6.1   Overview

The remainder of this book outlines methods for visual data mining, clustering, and predictive analytics. It also discusses how data mining is being used and describes a software application that can be used to get direct experience with the methods in the book.

1.6.2   Data Visualization

Visualization is a central part of exploratory data analysis. Data analysts use visualization to examine, scrutinize, and validate their analysis before they report their findings. Decision makers use visualization to explore and question the findings before they develop action plans. Each group of people using the data needs different graphics and visualization tools to do its work.

Producing high quality data graphics or creating interactive exploratory software requires an understanding of the design principles of graphics and user interfaces. Words, numbers, typography, color, and graphical shapes must be combined and embedded in an interactive system in particular ways to show the data simply, clearly, and honestly.

There are a variety of tables and data graphics for presenting quantitative data. These include histograms and box plots for displaying one variable (univariate data), scatterplots for displaying two variables (bivariate data), and a variety of multipanel graphics for displaying many variables (multivariate data). Visualization tools like dendrograms and cluster image maps provide views of data that has been clustered into groups. Finally, these tools become more powerful when they include advances from interactive visualization.

1.6.3 Clustering

Clustering is a commonly used approach for segmenting a data set into groups of related observations. It is used to understand the data set and to generate groups in situations where the primary objective of the analysis is segmentation. A critical component in any data clustering exercises is an assessment of the *distance* between two observations. Numerous methods exist for making this determination of distance. These methods are based on the type of data being clustered; that is, whether the data set contains continuous variables, binary variables, nonbinary categorical variables, or a mixture of these variable types. A series of distance calculations are described in detail in Chapter 3.

There are a number of approaches to forming groups of observations. *Hierarchical* approaches organize the individual observations based on their relationship to other observations and groups within the data set. There are different ways of generating this hierarchy based on the method in which observations and groups in the data are combined. The approach provides a detailed hierarchical outline of the relationships in the data, usually presented as a dendrogram. It also provides a flexible way of generating groups directly from this dendrogram. Despite its flexibility, hierarchical approaches are limited in the number of observations they are able to process, and the processing is often time consuming. *Partitioned*-based approaches are a faster method for identifying clusters; however, they do not hierarchically organize the data set. The number of clusters to generate must be known prior to clustering. An alternative method, referred to as *fuzzy* clustering, does not partition the data into mutually exclusive groups, as with a hierarchical or partitioned approach. Instead, all observations belong to all groups to varying degrees. A score is associated with each observation reflecting the degree to which the observation belongs in each group. Like partitioned-based methods, fuzzy clustering approaches require that the number of groups be set prior to clustering.

1.6.4 Predictive Analytics

The focus of many data mining projects is making predictions to support decisions. There are numerous approaches to building these models, and all can be customized to varying degrees. It is important to understand what types of models, as well as what parameter changes, improve or decrease the performance of the predictions. This assessment should account for how well the different models operate using data separate from the data used to build the model. Dividing the data into sets for building and testing the model is important, and common approaches are outlined in Chapter 4. Metrics for assessment of both regression and classification models are described.

Building models from the fewest number of independent variables is often ideal. Principal component analysis is one method to understand the contribution of a series of variables to the total variation in the data set. A number of popular classification and regression methods are described in Chapter 4, including multiple linear regression, discriminant analysis, logistic regression, and naive Bayes. Multiple linear regression identifies the linear relationship between a series of independent variables and a single response variable. Discriminant analysis is a classification approach that assigns observations to classes using the linear boundaries between the classes. Logistic regression can be used to build models where the response is a binary variable. In addition, the method calculates the probability that a response value is positive. Finally, naive Bayes is a classification approach that only works with categorical variables and it is particularly useful when applied to large data sets. These methods are described in detail, including an analysis of when they work best and what assumptions are required for each.

1.6.5 Applications

Data mining is being applied to a diverse range of applications and industries. Chapter 5 outlines a number of common uses for data mining, along with specific applications in the following industries: finance, insurance, retail, telecommunications, manufacturing, entertainment, government, and healthcare. A number of case studies are outlined and the process is described in more detail for two projects: a data set related to genes and a data set related to automobile loans. This chapter also outlines a number of approaches to data mining some commonly used nontabular sources, including text documents as well as chemicals. The chapter includes a description of how to extract information from this content, along with how to organize the content for decision-making.

1.6.6 Software

A software program called Traceis (available from http://www.makingsenseofdata. com/) has been created for use in combination with the descriptions of the various methods provided in the book. It is described in Appendix B. The software provides multiple tools for preparing the data, generating statistics, visualizing variables, and grouping observations, as well as building prediction models. The software can be used to gain hands-on experience on a range of data mining techniques in one package.

1.7 SUMMARY

The preceding chapter described a data mining process that includes the following steps:

1. *Definition*: This step includes defining the objectives of the exercise, the deliverables, the roles and responsibilities of the team members, and producing a plan to execute.

2. *Preparation*: The data set to be analyzed needs to be collected from potentially different sources. It is important to understand the content of the variables and define how the data will be used in the final analysis. The data should be cleaned and transformations applied that will improve the quality of the final results. Efforts should be made to reduce the number of variables in the set

TABLE 1.1   **Data Mining Tasks**

Type of task	Specific task	Description	Example methods
Descriptive	Association	Finding associations between multiple items of interest	Association rules, decision trees, data visualization
	Segmentation	Dividing a data set into groups that share common characteristics	Clustering, decision trees
	Outliers	Identifying usual observations	Clustering, data visualization
Predictive	Classification	A predictive model that predicts a categorical variable	Discriminant analysis, logistic regression, naive Bayes
	Regression	A predictive model that predicts a continuous variable	Multiple linear regression

to analyze. A subset of observations may also be needed to streamline the analysis.

3. *Analysis*: Based on an understanding of the problem and the data available, a series of data mining options should be investigated, such as those summarized in Table 1.1. Experiments to optimize the different approaches, through a variety of parameter settings and variable selections, should be investigated and the most promising one should be selected.

4. *Deployment*: Having implemented the analysis, carefully planning deployment to ensure the results are translated into benefits to the business is the final step.

1.8   FURTHER READING

A number of published process models outline the data mining steps, including CRISP_DM (http://www.crisp-dm.org/) and SEMMA (http://www.sas.com/technologies/analytics/datamining/miner/semma.html). In addition, a number of books discuss the data mining process further, including Shumueli et al. (2007) and Myatt (2007). The following resources provide more information on preparing a data set for data mining: Han and Kamber (2006), Refaat (2007), Pyle (1999, 2003), Dasu and Johnson (2003), Witten and Frank (2003), Hoaglin et al. (2000), and Shumueli et al. (2007). A discussion concerning technology standards for deployment of data mining applications can be found in Hornick et al. (2006).

CHAPTER 2

DATA VISUALIZATION

2.1 OVERVIEW

Data visualization is critical to understanding the content of the data. Data analysts use visualization to examine, scrutinize, and validate their analysis before they report their findings. People making decisions use visualization to explore and question the findings before they develop action plans. Each group needs different graphics and visualization tools to do its work.

Data sets often come from a file and are typically displayed as a table or spreadsheet of rows and columns. If the data set is small and all the data can be displayed on a single page, it can be analyzed or the results presented as a table. But as the number of rows (observations) and columns (variables) increase, long lists of numbers and statistical summarizations of them do not tell us all we need to know. Data graphics help us understand the context and the detail together. They help us think visually and provide a powerful way to reason about large data sets.

While data graphics are centuries old, the graphical user interfaces available today on every computer enable interactive visualization tools to be included in information software products. For example, online newspapers contain interactive graphics that allow the reader to interactively explore data, such as the demographics of voters in elections, or the candidates' source of income. Visualization tools for data sets with many variables, in particular, must display relationships of three or more variables on paper or display screens that are two-dimensional surfaces. An understanding of the basic design principles of data graphics and user interfaces will help to use and customize data graphics to support decision-making. This chapter reviews these principles.

Organizing graphics and visualization tools is not easy. They depict a variety of data types including numbers and categories. Different tools are used in different ways throughout the data analysis process: to look at summary statistics, examine the shapes of distributions, identify outliers, look for relationships, find groups of similar objects, and communicate results. They are used in different application areas to display, for example, the results of document searches in information retrieval or the correlation of patterns of gene expression with chemical structure activity in genomic research. The use of data visualization within software programs enable interactive techniques such as data brushing, which is the ability to simultaneously highlight the same data in several data graphics to allow an open-ended exploration of the data set.

Making Sense of Data II. By Glenn J. Myatt and Wayne P. Johnson
Copyright © 2009 John Wiley & Sons, Inc.

In this chapter, a section on the principles of graphics design and graph construction is initially presented. The next section looks at tables, an old and refined graphical form. The next three sections focus on graphical visualization tools for quantitative data. These tools are classified by the number of variables they display, and tools for one variable (*univariate* data), two variables (*bivariate* data), or many variables (*multivariate* data) are discussed. The sections on quantitative data are followed by a section on tools to visualize groups of observations. Finally, there is a section that discusses techniques for interacting with data visualizations to explore the data.

Data graphics and visualization tools can be easily found in print or on the internet. Many are examples of how *not* to communicate your statistical analysis or results, but some deserve careful study. Those commonly used by data analysts are included, in addition to some that are not well known but are effective in niche areas or written about by well-known statisticians and scientists: John Tukey, William Cleveland, Edward Tufte, Howard Wainer, and John Weinstein. These less well-known graphics illustrate certain design principles and provide examples from specific application areas of how easily visualization can reveal structure hidden within data or generate ideas for new designs.

2.2 VISUALIZATION DESIGN PRINCIPLES

Good design begins by asking who will use the results, why, and how? Since this chapter is about the visualization of data, it will focus only on those performing the analysis and the consumers of their analysis who are people making critical decisions. Data graphics help make arguments, but if essential details are left out, distorted, or hard to see, the consequences can be catastrophic. Before examining specific data graphics and visualization tools, the construction of a commonly used graph, the scatterplot will be reviewed. Some general principles will be described, along with the basics of graphics design.

2.2.1 General Principles

There are several general principles to keep in mind when designing data graphics.

Show the Data Edward Tufte emphasizes that "data graphics should draw the viewer's attention to the sense and substance of data, not to something else" (Tufte, 1983). The representations of the data through plots of symbols that represent values, categorical labels, lines, or shaded areas that show the change in the data, and the numbers on scales are what is important. The grids, tick marks on scales, reference lines that point out key events, legend keys, or explanatory text adjacent to outliers should never get in the way of seeing the data.

Simplify Choose the graphic that most efficiently communicates the information and draw it as simply as possible. You will know your drawing is done when you can take nothing more away—points, lines, words, symbols, shading, and grids— without losing information. For small data sets, tables or dot plots are preferable to

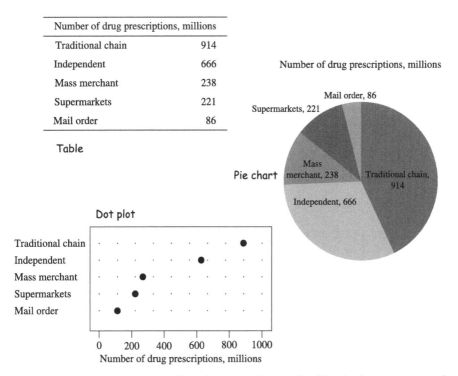

Number of drug prescriptions, millions	
Traditional chain	914
Independent	666
Mass merchant	238
Supermarkets	221
Mail order	86

Figure 2.1   Presenting data as a table, a pie chart, and a dot plot. The pie chart was generated by Microsoft Excel 2007. Source: NACDS Foundation Chain Pharmacy Industry Profile, Table 130, 2006

graphics. They are easier to understand and communicate the most information. Avoid what William Cleveland calls pop charts: pie charts, divided bar charts, and area charts that are widely used in mass media but carry little information (Cleveland, 1994). The same information is communicated in three ways in Fig. 2.1. Notice that the table, which displays the same information in a form more easily read and compared than the pie chart, takes up about half the space.

Reduce Clutter   Clutter comes from two sources. The first source is the marks on the drawing that simply crowd the space or obscure the data. If grid lines are needed at all, draw thin lines in a light shade of gray. Remove unnecessary tick marks. Look for redundant marks or shading representing the same number. For example, the height of the line and the use of the number above the bar in Fig. 2.2 restate the number 32.5. The second source of clutter is decorations and artistic embellishments.

Revise   Any good writer will tell you that the hard work of writing is rewriting. Graphic designers also revise to increase the amount of ink devoted to the data. The panels in Fig. 2.3 show the redesign of a scatterplot. In the second panel, we removed the grid. In the third panel, we removed unnecessary tick marks.

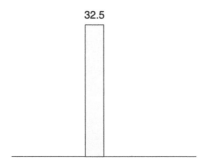

Figure 2.2 Histogram bar that redundantly encodes its height

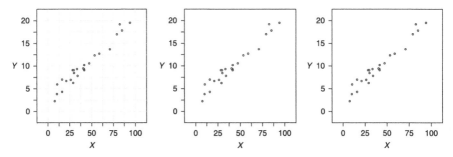

Figure 2.3 Revising graphs to show more data

Be Honest A graphic tells the truth when "the visual representation of the data is consistent with the numerical representation" (Tufte, 1998). Here are some ways that graphics can distort data:

- *Adjust the aspect ratio of the graph to overstate or understate trends.* The aspect ratio is the height of the data rectangle, the rectangle just inside the horizontal and vertical scales in which the points are plotted, divided by the width. By increasing or decreasing the height while keeping the width constant, one can make dramatic changes to the perceived slope of a line. In Fig. 2.4, note how much more the curve in the panel on the right appears to rise compared to the panel on the left.

- *Manipulate the scale.* This distortion is achieved through the use of a scale with irregular intervals. For example, consider the histograms in Fig. 2.5 of the income distribution, which shows the percentage of families with incomes in each class interval. In the panel on the left, a unit of the horizontal scale means two different things: a class interval size of $1,000 or a class interval size of $5,000. When the scale is corrected, as in the panel on the right, the percentage of families with incomes between $5,000 and $10,000 is now being fairly compared with the other class intervals. Another example of distortion is a large scale range that hides important variation in the data, as in Fig. 2.6.

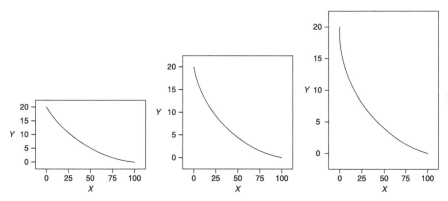

Figure 2.4   Making a more dramatic statement by adjusting only the aspect ratio

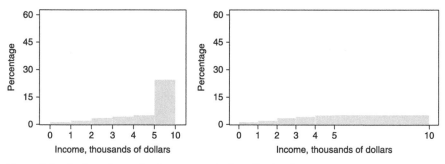

Figure 2.5   An irregular scale distorts magnitudes of unit values

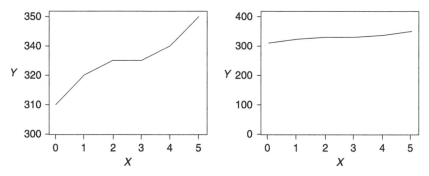

Figure 2.6   Increasing the range of a scale may hide important detail

2.2.2 Graphics Design

Data graphics show measured quantities and categories using a variety of graphical elements: points, lines, numbers, symbols, words, shading, and color. The designers of data graphics rely on knowledge from the field of graphics design. In this section the principles of page layout which help us structure the legend, scales, title, caption, and other parts of the graph are discussed. Discoveries about our visual systems that give us insight into ways to encode the data are examined. Finally, aspects of color and typography that are important in drawing graphics are discussed.

When designing the graph layout, it is important to be aware of the *visual hierarchy*, the *visual flow*, and the *grouping* of elements.

Visual Hierarchy Every page has a *visual hierarchy*. A visual hierarchy makes some content appear more important than other content. This can be done by: (1) moving the content to the upper-left corner; (2) separating different components with white space; (3) using larger and bolder fonts; (4) using contrasting foreground and background colors; (5) aligning it with other elements; and (6) indenting it, which logically moves the indented content to a deeper level in the hierarchy than the element above it.

Visual Flow Visual flow describes the path the eye follows as it scans the page. It is typical to read top-to-bottom and left-to-right (in Western cultures), but this can be controlled by creating focal points. Just as underlined words are used for emphasis, focal points are a graphical way of identifying what is important. Focal points attract the eye and the eye follows them from strongest to weakest. Some ways to create focal points include larger and bolder fonts, spots of contrasting color, and separation by white space.

Grouping Graphical elements are perceived as being part of a group when they are close together (*proximity*), have similar color or shading (*similarity*), are aligned along an invisible line or curve (*continuity*), or are positioned so that they appear to be within a closed form (*closure*). These Gestalt principles (named after the psychological theory which held that perception is influenced not only by the elements but also by context) can be applied to create a visual hierarchy or focal points in a graph without adding additional graphical elements. Figure 2.7 illustrates these four principles. In the top-left panel (proximity), although the shapes are irregularly sized, the eye sees two groups because the shapes in each group are close together with plenty of white space between the groups. In the top-right panel (similarity), the eye separates into two groups the shapes with similar color: the three light gray shapes and the two dark gray shapes. In the bottom-left panel (continuity), the eye separates the left group of shapes from the right by tracing the continuous edge along the left side of the right group and along the right side of the left group of shapes. In the bottom-right panel (closure), the eye traces the implicit rectangle that encloses the group of shapes in the right half of the panel.

In addition to the Gestalt principles which help us design layout, experimental psychologists have discovered other things about our visual systems that are useful in deciding how to graphically encode data values. Certain visual features such as color, texture, position and alignment, orientation, and size are processed almost instantaneously. These features are called *preattentive variables* and they give us options for encoding data so that we can find, compare, and group them without much mental effort.

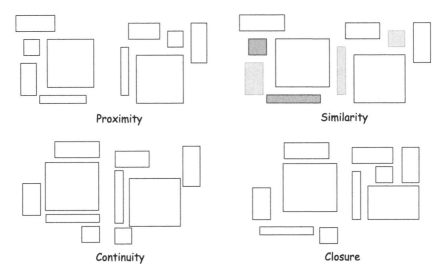

Figure 2.7   Illustrating Gestalt principles

Before illustrating preattentive variables, let us do two experiments that show preattentive processing in action. First, in Fig. 2.8, count the number of dark gray circles.

Now do the same in Fig. 2.9, which has twice as many circles.

This first experiment compared the search time it took you to find target circles encoded with the *color* preattentive variable in sets of circles. The time for each search should be more or less constant, because the searching is done in the brain's visual system in a preattentive stage when it is processing what you see.

In the second experiment shown in Fig. 2.10 of monotonous text, try to find all the numbers greater than or equal to 1.

Now try this again for Fig. 2.11.

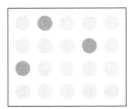

Figure 2.8   Graph with dark circles

Figure 2.9   Graph with additional dark circles

0.48	0.09	1.02	0.61	0.47	0.05	0.09	0.79	0.64	0.44
1.07	0.69	0.14	0.81	0.44	0.22	0.55	0.70	0.47	0.01
0.73	1.08	0.16	0.45	0.50	1.16	0.45	0.50	0.16	0.79
0.32	0.93	0.62	0.25	0.05	0.11	0.88	1.15	0.88	0.99
0.40	0.20	0.16	1.09	0.51	0.04	0.38	0.36	0.48	0.22

Figure 2.10   Graph of unencoded random numbers

0.48	0.09	**1.02**	0.61	0.47	0.05	0.09	0.79	0.64	0.44
1.07	0.69	0.14	0.81	0.44	0.22	0.55	0.70	0.47	0.01
0.73	**1.08**	0.16	0.45	0.50	**1.16**	0.45	0.50	0.16	0.79
0.32	0.93	0.62	0.25	0.05	0.11	0.88	**1.15**	0.88	0.99
0.40	0.20	0.16	**1.09**	0.51	0.04	0.38	0.36	0.48	0.22

Figure 2.11   Graph of random numbers that encodes critical numbers with font size and texture

The first experiment measured the time it took you to search unencoded data; the second measured the time to search encoded data. The search through monotonous text took longer because you needed to look at and consider each number. With the target text encoded as *texture* (bold font) and size (larger font), the search was almost instantaneous. Figure 2.12 graphically summarizes six preattentive variables.

Color   Color is used to encode the data, to shade the bars or plotting symbols, and to color the background of different parts of a graph, reference lines, or grids. The most important rule in choosing color is to never do anything that makes it impossible to read.

- *Use contrasting colors for foregrounds and backgrounds.* Light on dark or dark on light colors should be used. User interface designers use white backgrounds to indicate areas that can be edited. Dark backgrounds are rarely used because user interface controls such as text fields or buttons are usually not visually pleasing when overlaid on a dark background.

- *Never use red or green if these two colors must be compared.* People who are colorblind will not see the difference. Those colors affect about 10% of men and 1% of women.

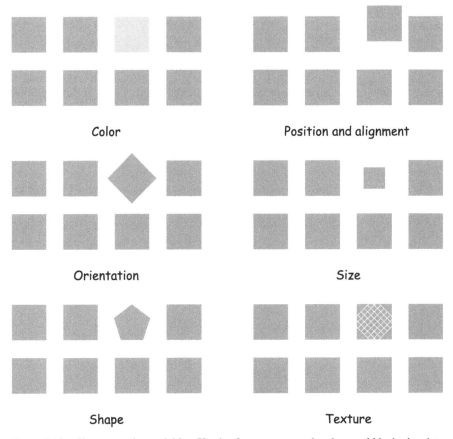

Figure 2.12   Six preattentive variables. If color figures were used, color would be broken into three variables: hue, brightness, and saturation

- *Never put small blue text on red or orange backgrounds or vice versa.* In fact, text cannot be read that is in a color complementary (on opposite sides of the color wheel) to the color of its background.
- *Use bold colors sparingly.* Bold colors, as well as highly saturated colors such as red, yellow, or green, tire the eye when you look at them for long periods of time. Use them sparingly. Light, muted colors are preferred for large areas like backgrounds.

Typography In graphs, small text is used for labels alongside tick marks, legend keys, and plotted symbols; normal text is used in titles and captions. As with color, choose fonts (the technical term is *typefaces*) that are easy to read. Text in small point sizes are easiest to read on computer displays when drawn in *sans-serif* fonts. Computer displays lack the resolution of the printed medium. In print, *serif* fonts look better. The more letters are differentiated from each other, the easier they are to read; therefore avoid using words in all caps except for headlines and short text.

2.2.3 Anatomy of a Graph

Show the data and *reduce clutter* are helpful principles but it helps to know how to apply them when drawing a graph. This section describes the graphical elements and the characteristics of a graph that clearly and succinctly show the important features of the data.

Figure 2.13 shows the annotated graph introduced by William Cleveland (Cleveland, 1994). It defines the terminology that will be used throughout this chapter. The following discussion explains each element along with aspects of good graph design. It starts with the data rectangle, the innermost element, and works outward to the edges of the graph.

Data Rectangle   This is the canvas on which data is plotted and the lines fitted to the data are drawn. It should be noted that the data rectangle seen in Fig. 2.13 is never

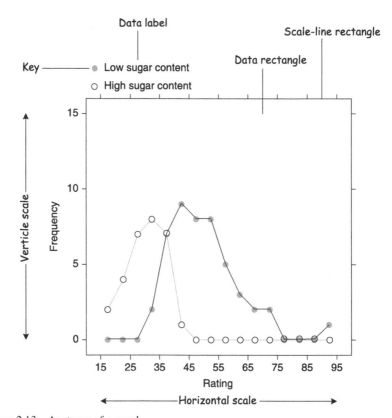

Figure 2.13   Anatomy of a graph

drawn but is only shown to identify the drawing area for plotting the graph. If the data values have labels that cannot be drawn adjacent to the plotted symbols without obscuring the data, consider using a different graphic. Reference lines and grids, if they are used at all, should be light and thin lines colored in a shade of gray that do not draw attention to themselves. If two data sets are superimposed in the data rectangle, it should be easy to visually separate the plotting symbols and connecting lines that belong to each set.

Plotting Symbol and Data Label   The choice of plotting symbol affects how conspicuous the point will be—especially if lines connect the points—and how easily points can be visually found and grouped into categories if different symbols have been used to encode a value's category as well as its magnitude. Filled circles make a good choice unless more than one datum has the same value and they will be plotted on top of each other. For this case, an unfilled circle can be combined with a jittering technique that randomly offsets each circle from its normal position to help single out data with the same value.

Scale-line Rectangle   The data rectangle and its surrounding margin is the scale-line rectangle, or everything just inside the frame. As discussed in the graphics design section above, white space is important for separation. The margins separate the data from the scales and keep the data points—particularly outliers in the corners or points that might otherwise fall on the horizontal and vertical scales—from getting lost. The data labels in the interior should not interfere with the quantitative data. Keys should be kept outside and usually above the frame; notes should be put in the caption or in the text outside this rectangle.

Reference Lines   To note important values, use a reference line or reference grid but do not allow it to interfere with the data. If the graph consists of multiple panels, be sure the line or grid is repeated in the same position in every panel.

Scales and Scale Labels   Choose the scales so that the data rectangle fills up as much of the scale-line rectangle as possible, but always allow for small margins. Zero need not be included on a scale showing magnitude. If the scale is logarithmic, make sure to mention it in the scale label. If the scales represent quantitative values, the horizontal scale, read left-to-right, should have lower values to the left of higher values; the vertical scale, read bottom-to-top, should have lower values below higher values.

When scatterplots are used to see if one variable is dependent on another, the graph is drawn in a certain way. By convention, the response or dependent variable is plotted on the vertical scale and the independent variable is plotted against the horizontal scale. Pairs of scale lines should be used for each variable. The vertical scale on the left should be reflected on the right; the horizontal scale below should be reflected above.

Sometimes there are large intervals with no data and it is necessary to break the scale to conserve space. This should be done by breaking the graph into separate panels that have a gap between; the numerical values on two sides of a break should not be connected.

It is sometimes useful to use two different scales—above and below, or to the left and the right—in order to visually index a data point by either scale. For example, the horizontal scale above might be a person's age while the one below might be the person's birth year.

Tick Marks and Labels   Tick marks should include the range of data. They should point outwards so that they do not encroach on the margin inside the scale-line rectangle and get entangled with data. A few tick marks and labels go a long way to providing a sense for the range and magnitude of the scale.

Key and Data Label   Sometimes when displaying multivariate data in scatterplots, values for two quantitative variables are encoded using the position of a symbol along the horizontal and vertical scales. If a third variable contains categorical data, each symbol may further encode that category's value using a different shape or color. The keys and associated data labels that explain which category is associated with which shape or color should be generated. The set of keys and their labels comprise the *legend*. Place legends outside but close to the scale-line rectangle, preferably above the rectangle since the eye, in a top-down visual flow, naturally stops first on the legend before it moves into the data rectangle.

Title and Caption   The title provides the headline. The caption explains and the explanation should be comprehensive and informative.

Aspect Ratio   The aspect ratio is not a graphical element you will find in Fig. 2.13 but it is an important topic in the design of graphs. The aspect ratio is the height of the *data rectangle*—not the scale-line rectangle—divided by the width.

The aspect ratio determines how well our eye can detect the rate of change of a curved line fitted to the data in a graph. Experiments in visual perception show that our ability to see changes as our eye moves along the line is greatest when the aspect ratio of the curve is "banked to 45°" (Cleveland, 1993). "Banking to 45°" is a technique that adjusts the aspect ratio in order to optimize your ability to perceive the changes in a curve. Like the banking of a road as it curves in a new direction, "banking" adjusts the aspect ratio so that subtle changes can be more easily seen as the eye travels along the fitted lines within. Figure 2.14 illustrates the idea, and further explanations can be found in William Cleveland's books (Cleveland, 1993, 1994).

Panels   A goal of graphing is to make comparisons. If the data are not too complex, this may be done by plotting the data points of several data sets in the data rectangle of the same panel. However, to avoid clutter, it may be necessary to split the graph into separate panels. In Fig. 2.15, the graph we started with has been split into two panels that show each set of points and their connected lines as *juxtaposed* panels rather than *superimposed* in one panel as in Fig. 2.13.

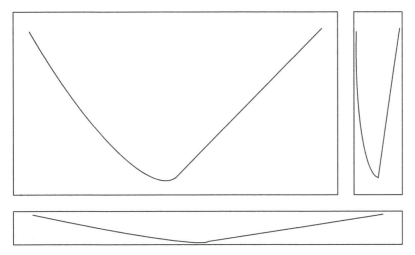

Figure 2.14   The top-left panel is banked to 45°

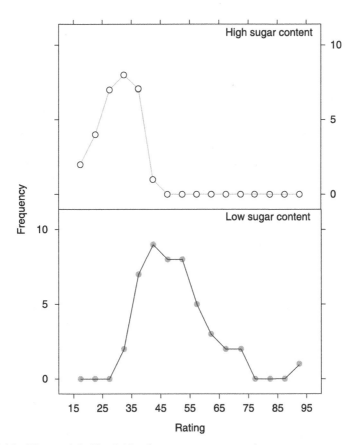

Figure 2.15   The graph in Fig. 2.13 redrawn as separate panels

Finally, it is important to keep in mind whether the purpose of the design is to communicate and illustrate results or explore the data. There are no rules that can be rigidly applied and it may take several iterations and some experimentation before getting it right.

2.3 TABLES

Tables are not graphics but they are effective in many situations: for showing exact numerical values, for small data sets, and "when the data presentation requires many localized comparisons" (Tufte, 1983).

2.3.1 Simple Tables

Simple data are often best presented by a simple table. However, especially when working with spreadsheets, the temptation is to reach for a pie chart to add to a Powerpoint presentation. Sorted tables communicate more information, more quickly, than pie charts. Pie charts force the viewer to find unaligned categorical labels inside or outside of irregularly spaced sectors within a circle, or decode colors of sectors using a legend. Carefully consider which alternatives best communicates what the data shows.

A good table should make the patterns and exceptions of the data obvious. It should summarize and explain the main features. Howard Wainer provides an example of how even a simple table can be transformed to reveal hidden structure in the data (Wainer, 2005). His example starts with a table of how nine Supreme Court justices voted on six important cases that was printed in *The New York Times* in July 2000. Both the rows and the columns in the table were sorted alphabetically by topic and justice. Except for Justice O'Connor, who voted with the majority on every case, no other pattern is apparent. Figure 2.16 shows the progression of rearranging a table of voting records of the Supreme Court justices.

Rearranging the columns and rows reveals something more in this data. To rearrange the table, start by cutting it into columns and then rearranging the columns in an order that gives highest priority to placing gray squares as far left as possible, starting with the top row and working to the bottom. The table will eventually look like the third panel in Fig. 2.16. Similarly, the table can be reassembled and cut into rows. Repeating the same procedure as for the columns, rearrange the rows in an order that gives highest priority to placing gray squares as high as possible, starting with the left column and working to the right. The last panel in Fig. 2.16 makes the pattern and exception obvious. Two groups of justices tend to vote together on certain kinds of cases. The exception is O'Connor who, as the swing justice, votes with the majority.

Something else to notice in the final table of Fig. 2.16 is that it is no longer sorted alphabetically. Ordering a table alphabetically may provide an index to the table, but almost always reveals nothing more. Even for a simple table with labels in the first column, the table should be ordered by the variable with the most important data values, as in Fig. 2.17.

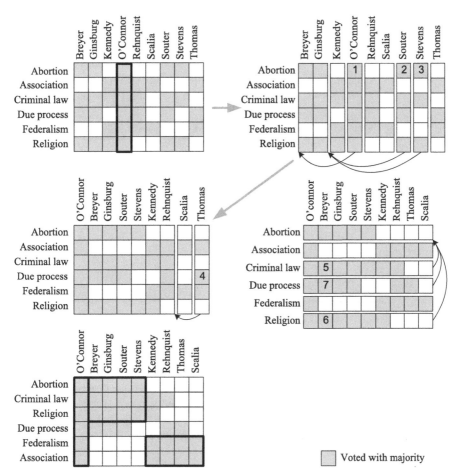

Figure 2.16   Rearranging the columns and rows of a table of U.S. Supreme Court justice rulings reveals groups of judicial orientation

Group B	26%
Group A	32%
Group C	42%

Figure 2.17   A table ordered by its most important content, not alphabetically

2.3.2 Summary Tables

A summary table displays summary statistics for observations that have been grouped by a single variable. A common format for a summary table is shown in Table 2.1.

Each row is a group of observations. The variable used to group observations in the table is in the first column; the number of observations grouped, or the count, is in the second column. Each remaining column contains the values of a summary statistic

TABLE 2.1   Format for a Summary Table

Variable a	Count	Variable x summary	Variable y summary	...
a_1	Count (a_1)	Statistic (x) for group a_1	Statistic (y) for group a_1	...
a_2	Count (a_2)	Statistic (x) for group a_2	Statistic (y) for group a_2	...
a_3	Count (a_3)	Statistic (x) for group a_3	Statistic (y) for group a_3	...
...	...	...	...	...
a_n	Count (a_n)	Statistic (x) for group a_n	Statistic (y) for group a_n	...

TABLE 2.2   Summary Table Showing Average *mpg* for Different Cylinder Vehicles

Cylinders	Count	Mean, *mpg*
3.0	4	20.55
4.0	199	29.28
5.0	3	27.37
6.0	83	19.97
8.0	103	14.96

applied to each group of observations for some variable in the data set. Common statistics used to summarize the values of a variable for the observations in a group are the *mean, median, sum, minimum, maximum,* or *standard deviation.* The value of a cell—for example, *Statistic (x) for group a_1* in Table 2.1—is the descriptive statistic calculated over the values in the group, for the selected variable. Table 2.2 summarizes the *mpg* for vehicles (observations) grouped by the number of cylinders in the vehicle's engine. Summary tables are discussed in more detail in the first book in this series (Myatt, 2007).

2.3.3   Two-Way Contingency Tables

It is often important to view how frequencies of observations are distributed across the categories or ranges of numeric values in two variables. For example, suppose two variables are being analyzed in a data set of car types: number of cylinders and miles per gallon. You may want to know how many types of cars with four cylinders travel less than 30 miles per gallon of gas or how many types of cars with more than four cylinders meet the new 35 miles per gallon standards. A *two-way contingency table* helps answer these kinds of questions. The general format of a contingency table is shown in Table 2.3 and a contingency table for this example in Table 2.4. Two-way contingency tables are discussed in more detail in the first book of this series (Myatt, 2007).

2.3.4   Supertables

Elaborate, carefully designed tables that Edward Tufte calls *supertables* can summarize and show detail, and be as engaging as a well-written news article. A good example

TABLE 2.3   Contingency Table Format

		Variable x		Totals
		Value 1	Value 2	
Variable y	Value 1	$Count_{11}$	$Count_{21}$	$Count_{1+}$
	Value 2	$Count_{12}$	$Count_{22}$	$Count_{2+}$
		$Count_{+1}$	$Count_{+2}$	Total count

TABLE 2.4   Contingency Table Summarizing Counts of Cars Based on the Number of Cylinders and Ranges of Fuel Efficiency (mpg)

	Cylinders = 3	Cylinders = 4	Cylinders = 5	Cylinders = 6	Cylinders = 8	Totals
mpg (5.0–10.0)	0	0	0	0	1	1
mpg (10.0–15.0)	0	0	0	0	52	52
mpg (15.0–20.0)	2	4	0	47	45	98
mpg (20.0–25.0)	2	39	1	29	4	75
mpg (25.0–30.0)	0	70	1	4	1	76
mpg (30.0–35.0)	0	53	0	2	0	55
mpg (35.0–40.0)	0	25	1	1	0	27
mpg (40.0–45.0)	0	7	0	0	0	7
mpg (45.0–50.0)	0	1	0	0	0	1
Totals	4	199	3	83	103	392

How different groups voted for president				
	Carter	Reagan	Anderson	Carter–Ford in 1976
Democrats (47%)	66	26	6	77–22
Independents (23%)	30	54	12	43–54
Republicans (11%)	11	84	4	9–90
Liberals (17%)	57	27	11	70–26
Moderates (46%)	42	48	8	51–48
Conservatives (28%)	23	71	4	29–70
Family income				
Less than $10,000 (13%)	50	41	6	58–40
$10,000–$14,999 (14%)	47	42	8	55–43
$15,000–$24,999 (30%)	38	53	7	48–50
$25,000–$50,000 (32%)	32	58	8	36–62
Over $50,000 (5%)	25	65	8	—
Professional or manager (40%)	33	56	9	41–57
Clerical, sales or other				
white–collar (11%)	42	48	8	46–53
Blue–collar worker (17%)	48	47	5	57–41
Agriculture (3%)	29	66	3	—
Looking for work (3%)	55	35	7	65–34

Figure 2.18   Portion of supertable showing voter profiles for the 1976 and 1980 U.S. elections

of this is a table designed by Tufte for *The New York Times* that shows a profile of voters in the 1976 and 1980 U.S. presidential elections. Portions of this table are shown in Fig. 2.18 (Tufte, 1983). In the original table, 410 numbers are shown in 20 clusters of tabular paragraphs. You can learn how the vote split across various demographic categories and how the voting patterns changed between 1976 and 1980 by reading across the line or down within each cluster of three to seven lines.

2.4 UNIVARIATE DATA VISUALIZATION

2.4.1 Bar Chart

A *bar chart* is simply a graphical way to quickly make qualitative comparisons of a set of values by drawing each value in the set as a rectangular bar with a length proportional to the value it represents. A bar twice as long has twice the value; a bar just a little longer than the next one represents a value just a little higher.

The bar chart in Fig. 2.19 shows part of a profile from sections of a congressional report on the membership of the 110th U.S. Congress. A member of congress—senator, representative, or delegate—is an observation. Each bar represents

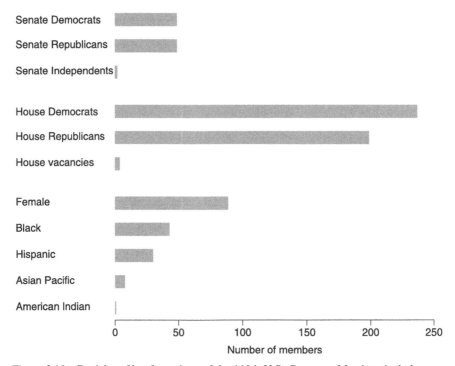

Figure 2.19 Partial profile of members of the 110th U.S. Congress. Members include senators, representatives, and delegates. Source: CRS Report for Congress, #RS22555; http://www.senate.gov/reference/resources/pdf/RS22555.pdf

the number of members in a specific category. Related categories are grouped and separated with white space. For a general bar chart, the bars may be presented in any order. The bars are ordered within each group from most to least members. They are drawn horizontally to make it easy to read the categories labeling the bars. Only six tick marks are included at intervals of 50, which makes for round numbers with few digits. The tick marks are far enough apart so that the white reference lines that cut through the longer bars can be seen distinctly and aligned with the horizontal scale.

With a quick scan of the chart, a number of facts can be easily deduced, such as that the independents break ties in the senate, the Democrats have a sizeable majority in the house, there are just under 200 house Republicans, and there are about twice as many women as African Americans.

2.4.2 Histograms

Histograms, when used for presentation, address questions like how many different car models could I choose to buy that travel over 30 miles per gallon of gas? Or how many inefficient models are being offered compared with highly efficient models? Histograms help answer these kinds of questions by displaying a *frequency distribution* of a data set's variables. Before histograms are discussed, it will be necessary to describe frequency distributions and how to create them.

A frequency distribution groups the data values into *classes* or *categories*. The number of observations in each class is the *class frequency*—a count of how many observations are in the class. A class for all but continuous variables is easy to identify. It is a category for nominal or ordinal variables, such as product names or the list *low, medium, high*. It is an integer for discrete quantitative variables, such as 2, 3 or 4 for the variable *family size*. But for continuous variables, a class means something different.

Consider a data set of cars with the variable *mpg* that has values from 9.0, the smallest, to 44.6, the largest. The difference between the smallest and the largest is the *range*. Some of the values within the range are spread out but others are tightly clustered around a specific value, such as 32.0, 32.1, 32.2, 32.4, 32.7, 32.8, and 32.9. Trying to display each of these as separate bars in a histogram would not be effective, since seven bars of almost imperceptible height would have to be placed side-by-side. This is solved by dividing the range into *class intervals* and assigning each value that falls within the interval to that class.

The number of class intervals you choose affects the width of the interval and, when the histogram is drawn, the length of the bars. With too many intervals the bars become so narrow that the histogram's contour may be too ragged to see. Fewer intervals lose some detail but the wider bars make it easier to see the histogram's shape. You can see the differences in Fig. 2.20.

How many to use is determined largely through trial and error by watching how the histogram changes shape after each adjustment. Consider starting with somewhere between 5 and 20 intervals, depending on the data set. When choosing the number of intervals, also consider the endpoint values. For the *mpg* variable, nine were selected so that the labels at the scale's tick marks, which mark the boundaries of each

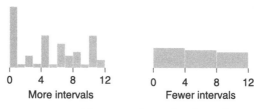

Figure 2.20 Histogram contours with more or fewer class intervals

interval, were divisible by 5. It is easy to read 5, 10, 15, ..., 50; less so if the labels are 9, 13, 17, ..., 45.

Once the number of intervals has been determined, along with the endpoint values, you need to adopt an endpoint convention for values that fall on the boundary between two intervals. In Fig. 2.21, to which class should 15.0 be added: the "10–15" or the "15–20" class? The choice depends on the selected endpoint convention. For endpoint conventions where the left endpoint, for example, 10, is included in the interval but the right endpoint, 15, is excluded, you add the observation to the "15–20" class.

There is a family of histograms that are commonly used: (1) the frequency histogram, (2) the relative frequency histogram, (3) the density histogram, and (4) the cumulative frequency histogram.

The *frequency histogram* displays the frequency of each class. The height of the bar represents the size of each class and is an absolute number. The vertical scale shows the count of the number of observations. Figure 2.22 shows frequency histograms for categorical and continuous variables.

The *relative frequency histogram* in Fig. 2.23 shows the fraction of times the values in the class appear. The height of the bar represents the ratio of the class size to the total size of the set, where the vertical scale is the relative frequency, that is a number from 0 to 1.

$$\text{relative frequency of class} = \text{class size}/\text{size of total set}$$

The sum of the heights, or the relative frequencies, of all classes in this histogram is 1 ($0.1 + 0.2 + 0.4 + 0.3 = 1$).

The *density histogram* in Fig. 2.24, closely related to the relative frequency, allows comparison of unit distributions across class intervals that vary in width (Freedman et al., 1998). The *height* of a bar represents the density, or how many

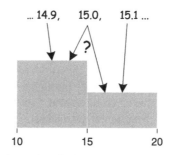

Figure 2.21 Effect of class interval endpoint convention

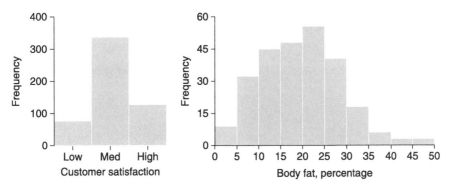

Figure 2.22   Frequency histograms of categorical and continuous variables

observations are in each unit of the histogram within the class represented by the bar. The *area* of the bar—its density per unit width times the width of the interval— represents the relative frequency of the class. The vertical scale is a density scale.

$$\text{density of class} = \frac{\text{relative frequency}}{\text{width of class}} = \frac{\text{class size}}{\text{class width} \times \text{size of total set}}$$

The sum of the *areas* is 1:

$$(0.02 \times 5) + (0.04 \times 5) + (0.07 \times 10) = 0.1 + 0.2 + 0.7 = 1$$

The *cumulative frequency histogram* shows the frequency of elements below a certain value. In this histogram, each bar represents the cumulative count of its class size added to the class sizes of all classes with smaller values. In Fig. 2.25, the income of 70 of 100 units, people or families, is less than \$15,000. The size of the last class equals the count of all observations in the set.

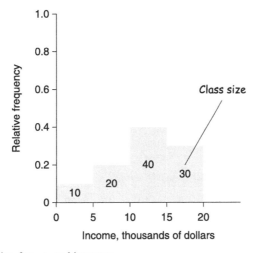

Figure 2.23   Relative frequency histogram

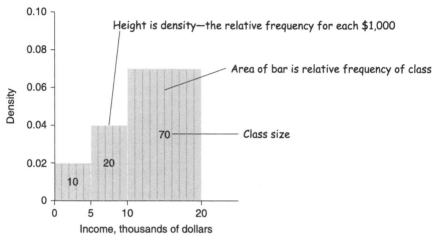

Figure 2.24   Density histogram

While a histogram helps see the frequency of the categories for a nominal or ordinal variable, it is most helpful for discrete and continuous quantitative variables, such as *family size* or *income*. During data exploration and analysis it is important to see other aspects of the frequency distribution and descriptive statistics of the variable: its shape, the location of its center, whether it is skewed toward one side or the other, peaks in the distribution, and any outliers. The first book of the series, in the section on descriptive statistics, shows how the frequency histogram is used for this kind of analysis (Myatt, 2007).

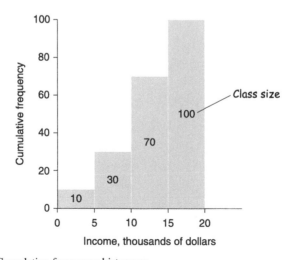

Figure 2.25   Cumulative frequency histogram

2.4.3 Frequency Polygram

A *frequency polygram* displays identical information to the frequency histogram. The histogram bars are replaced by connected points, as in Fig. 2.26.

2.4.4 Box Plots

The *box plot* is an analyst's tool used only for interval and ratio variables that graphically displays a five-number summary of a batch of numbers. To illustrate the box plot, we use a data set, shown in Table 2.5, of the population of the 21 largest metropolitan statistical areas of the United States in 2006. It is taken from a spreadsheet downloaded from the U.S. Census Bureau.

In Fig. 2.27, an annotated box plot, shown in the middle of the figure, summarizes the population of the largest metropolitan areas. Above the box plot is a list of the values ordered from smallest to largest. In the list of numbers at the top of the graphic, the important five values, highlighted in bold, are labeled and indexed to mark the corresponding vertical bars in the box plot.

Below the plot is part of the horizontal scale on which the population sizes have been plotted. Ordinarily the horizontal scale would lie empty below the box plot, but here the plotted values help to understand how the shape of the box, the position of the vertical bars, and the length of its extensions give the analyst a feel for the features of the data.

The five vertical bars mark the location on the horizontal scale of key markers in the ordered set of values for the variable. The two that lie on the outermost fringes mark the location of the smallest and the largest values. The other three form part of a box that defines the region containing the central half, or midspread, of the data. This region is known as the *fourth-spread* because it contains the lower and upper fourths. The width of the box shows the *spread*. The crossbar of the box, or the *median*, is a measure of the location of the center of the distribution. The position of the median relative to the lower and upper quartiles gives an indication of *skewness*: the symmetry or balance of the distribution curve around the center. In our example, the median's position to the right indicates that those data

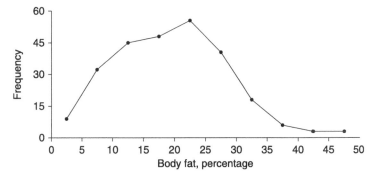

Figure 2.26   Frequency polygram showing the distribution of body fat

TABLE 2.5 Population of the 21 Largest Metropolitan Statistical Areas in the United States

New York–Northern New Jersey–Long Island, NY–NJ–PA	18,818,536
Los Angeles–Long Beach–Santa Ana, CA	12,950,129
Chicago–Naperville–Joliet, IL–IN–WI	9,505,748
Dallas–Fort Worth–Arlington, TX	6,003,967
Philadelphia–Camden–Wilmington, PA–NJ–DE–MD	5,826,742
Houston–Sugar Land–Baytown, TX	5,539,949
Miami–Fort Lauderdale–Miami Beach, FL	5,463,857
Washington–Arlington–Alexandria, DC–VA–MD–WV	5,290,400
Atlanta–Sandy Springs–Marietta, GA	5,138,223
Detroit–Warren–Livonia, MI	4,468,966
Boston–Cambridge–Quincy, MA–NH	4,455,217
San Francisco–Oakland–Fremont, CA	4,180,027
Phoenix–Mesa–Scottsdale, AZ	4,039,182
Riverside–San Bernardino–Ontario, CA	4,026,135
Seattle–Tacoma–Bellevue, WA	3,263,497
Minneapolis–St Paul–Bloomington, MN–WI	3,175,041
San Diego–Carlsbad–San Marcos, CA	2,941,454
St Louis, MO–IL	2,796,368
Tampa–St Petersburg–Clearwater, FL	2,697,731
Baltimore–Towson, MD	2,658,405
Denver–Aurora, CO	2,408,750

Source: http://www.census.gov/population/www/estimates/metro_general/2006/CBSA-EST2006-02.xls

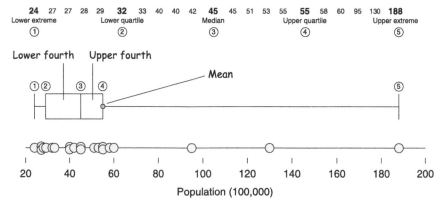

Figure 2.27   An annotated box plot of the 21 largest metropolitan statistical areas in the United States in 2006. Source: http://www.census.gov/compendia/statab/tables/08s0020.xls

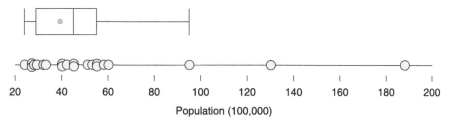

Figure 2.28   The revised box plot after imposing outlier cut-off limits

values in the upper fourth are gathered together more so than the ones in the lower fourth, moving them upwards.

The horizontal lines that stretch toward the horizons on the left and right show how the tails of the distribution curve taper off. In our example, the horizontal lines extend to the smallest and largest value; in more advanced analysis, *outlier cutoffs* may be imposed to limit their reach. Sometimes, values too far out throw off statistical calculations. If so, it is helpful to define outlier cutoff limits. For example, one definition might be to use the closest values on either end just within the cutoffs, where the cutoffs are defined as:

$$\text{upper outlier cutoff} = \text{upper quartile} + (1.5 \times \text{box width})$$
$$\text{lower outlier cutoff} = \text{lower quartile} - (1.5 \times \text{box width})$$

The revised box plot based on that definition for our example would be redrawn as shown in Fig. 2.28.

Notice that although we excluded two values from the box plot range, the position of the box and the median did not change. That is because the fourths and the median resist the impact of the outliers. In contrast, the dot, which represents the mean, shifted far to the left because the recalculation of the mean was significantly influenced by just two elements, New York and Los Angeles, which are no longer included in the recalculation.

2.4.5   Dot Plot

When the measurements of a quantitative variable are labeled and the data set is small, a *dot plot* displays the labeled data better than bar charts, divided bar charts, or pie charts. Not only are the long bars of a bar chart visually imposing, but they only work when the baseline of the graph is zero, otherwise the length of the bar is meaningless. Because the labels and scales of a dot plot flow horizontally, they can easily be read from left-to-right. If the rows are ordered by the values rather than alphabetically, patterns and trends can be identified. Figure 2.29 shows the population of the fifteen largest metropolitan statistical areas from Table 2.5.

There are other ways to extend dot plots. For example, to compare the population from the previous decade, superimpose different plotting symbols to show the values from each set.

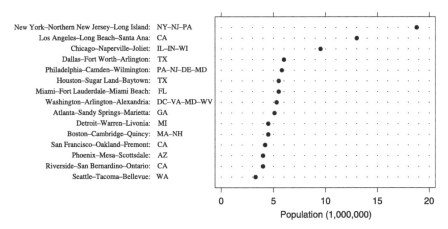

Figure 2.29   A dot plot of the 15 largest metropolitan statistical areas in the United States

2.4.6   Stem-and-Leaf Plot

The stem-and-leaf plot displays a histogram of a batch of numbers using digits in the numbers to form the tick mark label and the bar. By splitting the digits of each number into a stem and a leaf, and using the leaf—a single digit—to help draw the histogram bar, the numeric values can be seen along with their distribution. No information is lost. It is used during exploratory analysis to see the batch of numbers as a whole and to notice the same features: symmetry, spread, outliers, concentrations, and gaps.

The construction of the plot is best explained with an example. The data in Table 2.6 is extracted from a spreadsheet produced by the FBI and shows homicide rates in 2005 by state, ordered by homicide rate.

TABLE 2.6   Homicide Rates per 100,000 by U.S. State in 2005, Ordered by Rate

North Dakota	**1.1**	Massachusetts	2.7	New Jersey	4.8	Arizona	6.9
Iowa	1.3	Wyoming	2.7	Florida	5.0	California	6.9
Vermont	1.3	Colorado	2.9	Ohio	5.1	Missouri	6.9
Maine	1.4	Connecticut	2.9	Oklahoma	5.3	Tennessee	7.2
New Hampshire	1.4	Rhode Island	3.2	Indiana	5.7	Mississippi	7.3
Hawaii	1.9	Washington	3.3	Illinois	6.0	New Mexico	7.4
Montana	1.9	Wisconsin	3.5	Michigan	6.1	South Carolina	7.4
Minnesota	2.2	Kansas	3.7	Pennsylvania	6.1	Alabama	8.2
Oregon	2.2	Delaware	4.4	Virginia	6.1	Nevada	8.5
South Dakota	2.3	West	4.4	Georgia	6.2	**Louisiana**	**9.9**
Utah	2.3	New York	4.5	Texas	6.2	**Maryland**	**9.9**
Idaho	2.4	Kentucky	4.6	Arkansas	6.7		
Nebraska	2.5	Alaska	4.8	North Carolina	6.7		

Source: http://www.fbi.gov/ucr/05cius/data/table_05.html

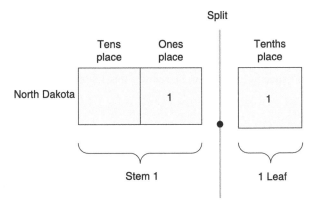

Figure 2.30   Splitting the values of homicide rates between the ones-place and tenths-place

First start by noting the lowest and highest values: 1.1 in North Dakota and 9.9 in Louisiana and Maryland. To create the stem-and-leaf plot, each value between 1.1 and 9.9 will be split into two parts: a stem and a leaf. The stem will become the base of the histogram and each leaf will be added to the histogram bar. But, looking at the numbers as strings of digits, between which *place-values* should the numbers be split? The digits to the left of the selected place-value will become the *stem*; the digit, just one, to the right becomes the *leaf*. It should be noted that, by choosing to have more than one digit to the right of the split, the leaf number should be rounded so that only one digit remains. In our example, values between the ones and the tenths place-values are split, as shown in Fig. 2.30.

Next create the bases for the histogram bars by writing all integers between the lowest and highest stem numbers to the left of the splitting bar as shown in the left panel of Fig. 2.31. Then traverse the list of states, adding each leaf digit to its corresponding stem. So for North Dakota, add 1 next to stem 1; for Iowa, add 3 next to stem 1; for Vermont, add 3 next to stem 1; and so on. Since the homicide rates were ordered, the leaf digits will naturally be ordered from 0 to 9 as they are added

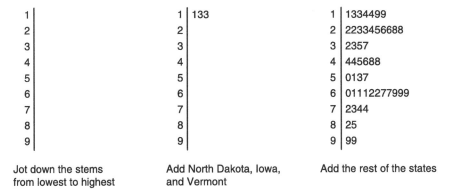

Figure 2.31   Constructing a stem-and-leaf plot

to the histogram bar. The middle panel of Fig. 2.31 shows the results after adding the homicide rates for North Dakota, Iowa, and Vermont; the right panel shows the results after adding all 50 states. The result is a histogram where digits are both content and building material for the histogram bars. The shape of the homicide rates for all 50 states and specific values can be seen in this graphic.

2.4.7 Quantile Plot

The *quantile plot*, or q-plot, helps visualize the distribution of a variable through the use of a scatterplot. Scatterplots are discussed in the next section. q-plots also set the stage for quantile–quantile plots, or q–q plots, used to compare distributions of different groups of the same variable or of two different variables.

We illustrate the q-plot by looking at a variable, X, of 15 randomly generated numbers, as seen in Table 2.7. The values of X are ordered from smallest to largest.

One way to see the distribution would be to plot each value on the vertical scale against some function of the index of X on the horizontal scale. In Fig. 2.32, two scales are shown. The first scale is based on i, where i ranges from 1 to N and N is the size of the variable, or 15 in our example. However, using the index for the horizontal scale will not enable the distributions of two variables to be compared.

Instead of a scale based on the index values, a normalized scale called an *f-value* scale is created. An *f quantile* of a distribution is a number, q, where approximately a fraction f of the values of the distribution is less than or equal to q; f is the *f-value* of q. *f*-values based on i are calculated by the following equation:

$$f_i = (i - 0.5)/N$$

In this example the equation is:

$$f_i = (i - 0.5)/15$$

When *f*-values other than f_i are needed, it is necessary to linearly interpolate and extrapolate the X_i and f_i values. Figure 2.33 magnifies the lower left corner of Fig. 2.32.

TABLE 2.7 A Table of the Indices and the Corresponding Values of a Variable X of Random Numbers

i	X_i	i	X_i	i	X_i
1	0.6	6	4.2	11	5.8
2	1.1	7	4.8	12	6.6
3	2.6	8	5.3	13	8.4
4	2.6	9	5.5	14	8.6
5	4.0	10	5.7	15	9.5

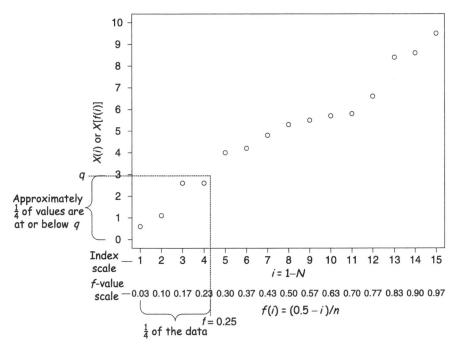

Figure 2.32   Quantile plot showing the index and f-value scales on the horizontal scale

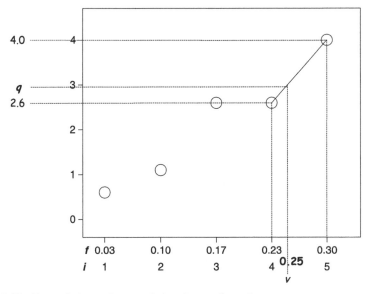

Figure 2.33   Interpolating and extrapolating the v and q values

Notice that 0.25 has no f_i-value because it falls between f_4 and f_5. A value of v is required such that:

$$(v - 0.5)/n = f$$

Solving for v results in:

$$v = nf + 0.5$$

To compute v in this example:

$$v = (15)(0.25) + 0.5 = 4.25$$

Now it is necessary to linearly extrapolate to find the quantile q on the vertical scale. Let k be the integer part of v and p be the fractional part. For this example k is 4 and p is 0.25. It is known that q will fall between X_k and X_{k+1}. The following general equation uses the integer and fractional parts of v to compute q:

$$q = (1 - p)X_k + pX_{k+1}$$

In this example, q falls between X_4 and X_5, k is 4, and p is 0.25. The following equation solves for q:

$$q = (1 - 0.25)X_4 + 0.25X_5 = (0.75)(2.6) + (0.25)(4.0) = 2.95$$

So the f-value 0.25 has the quantile value 2.95. For more details see William Cleveland's books on graphing and visualizing data (Cleveland, 1993, 1994).

2.4.8   Quantile–Quantile Plot

The quantile–quantile plot, or q–q plot, is used to compare distributions. This is done by plotting the quantiles of each variable or group against each other in a scatterplot. To illustrate a q–q plot, two plots are compared that use randomly generated numbers. To keep the example simple, both variables are of equal size. It becomes more complicated if one set is larger and the details for handling the general case can be found in William Cleveland's book on graphing data (Cleveland, 1994).

The first set of 15 random numbers was shown in Table 2.7 and plotted in Fig. 2.32. The second set is shown in Table 2.8.

TABLE 2.8   A Table of the Indices of Y and the Corresponding Values of 15 Randomly Generated Values Sorted from Smallest to Largest

i	Y_i	i	Y_i	i	Y_i
1	2.4	6	5.6	11	7.2
2	3.5	7	5.7	12	7.5
3	3.5	8	5.8	13	8.3
4	4.0	9	6.4	14	8.7
5	5.0	10	6.8	15	9.3

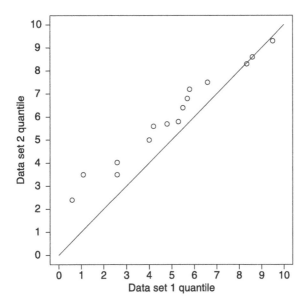

Figure 2.34   A q–q plot graphing the quantiles of 15 randomly generated numbers from data sets 1 and 2

Figure 2.34 graphs the quantiles of numbers from two columns of 15 randomly generated numbers in an Excel spreadsheet. To plot this graph, plot the value (X_1, Y_1), then (X_2, Y_2), and so on through (X_{15}, Y_{15}). A reference line with a slope of 1 has been drawn from (0, 0) to (10, 10) to allow comparison against the ideal. If one found that the points that lay on this line were evenly spread along it, it would likely indicate that the random number generator was not random.

Quantile–quantile plots are also used to compare the distribution of a variable against the normal distribution. By convention, the normal distribution is plotted on the horizontal scale and the variable's distribution is plotted on the vertical scale.

2.5   BIVARIATE DATA VISUALIZATION

2.5.1   Scatterplot

Scatterplots are widely used for comparing one variable with another. The first book in the series described scatterplots and showed how they provide a first look at the data to see if there are *linear* or *nonlinear* relationships between two continuous variables, or if there is any relationship at all. The patterns of the plotted symbols on the graph leave an impression of *correlation*. A computed correlation coefficient communicates information about the linear relationship of two variables: the direction of the relationship and how strongly they relate. Figure 2.35 shows some examples.

When the pattern of points moves upward from lower-left to upper-right, the correlation is positive; when the pattern of points moves downward from upper-left

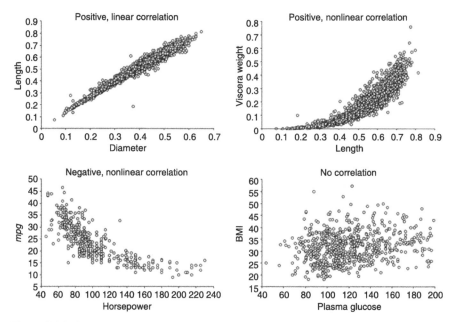

Figure 2.35   Examples showing correlation relationships of two variables

to lower-right, it is negative. When the points move more or less along an imagined straight line, the variables likely have a *linear* relationship; if not, the variables may have a *nonlinear* relationship defined by a more complex mathematical function, or have no relationship at all. When the points cluster tightly around an imagined line, the correlation is likely closer to 1.0 or −1.0; the more they spread out, the closer to 0 the correlation is likely to be.

2.6   MULTIVARIATE DATA VISUALIZATION

It is often desirable to visualize a data set with more than two variables, or *multivariate* data. The graphic that represents this data must still be drawn in print or on computer display in two dimensions. Somehow the complexity of many dimensions must be condensed to just two. There are a variety of ways to do this and they are mentioned in this introduction even though most will not be covered in this book.

First snapshots within multidimensional space can be taken and organized into a matrix of plots so that many pictures can be seen at one time. Multiple panels, such as those found in histogram or scatterplot matrices, organize these snapshots effectively. This section focuses primarily on tools like these.

A second way is to add to the two dimensions of the display or writing surface encodings of more dimensions using preattentive variables, such as color or texture. For example, a plotted point on a graph which has a position on a horizontal and vertical scale might also use several different symbols and color to allow it to encode two more categorical variables. This is a straightforward application of the design principles discussed in the section above.

A third way, particularly if working with three variables, is to encode the values in a three-dimensional cartesian space with x, y, and z coordinates and leave the responsibility of drawing the three-dimensional shapes and surfaces to software. Three-dimensional visualization is a complex subject with its own issues and it is not covered in this book.

Fourth, a highly dimensional space can be flattened into two dimensions in ways that preserve geodesic distances—like a cartographer who takes the Earth's surface and flattens it into maps for an atlas or contour maps for hikers. Methods such as multidimensional scaling can be used to create two-dimensional maps that maintain some abstract notion of distance: They use mappings from the high to low dimensional space where observations that were close in high dimension space remain close in low dimension space, and observations that were far away are still far away. These methods also are beyond the scope of this book.

Finally, there are creative ways to encode several dimensions. The most famous of these is an abstract map drawn by Charles Joseph Minard (1781–1870) of Napolean's Russian campaign of 1812 (Tufte, 1983). His drawing encodes six variables: the position of the troops—*latitude* and *longitude* in Cartesian coordinates—and troop *size* as it advanced and retreated, *distance* between geographical areas, the *temperature* throughout the march, and *dates* of important events. However, graphics like these begin to drift from data visualization into information visualization and are not covered here.

The focus of this section is on methods to capture and see the structure of spaces beyond two dimensions. To visualize the breadth and depth of the data's structure, a handful of widely used multipanel tools use techniques that *juxtapose* multiple panels by placing them side-by-side, or *superpose* data by integrating them into the same panel.

Before discussing each tool, two design issues relevant to multipanel displays are considered: how to organize the panels and how to compare across panels to see subtle variation and changes between them.

Organizing Panels in a Matrix The matrix, a two-dimensional structure, is the container for panels in multipanel displays. Sometimes panels are created that have no relationship to each other. They are like photographs from different scenes. They may be arranged in any order without impacting our understanding of the content. At other times, multipanel displays are created where the panels represent subsets that have been grouped by some ordered variable. They are like a series of related photographs taken of a panoramic view. In these cases, the convention is to use the columns and rows as a coordinate system and to place the panels in order from left-to-right and from bottom-to-top. These panels will be referred to by column and row number. In Fig. 2.36, panel (1, 2) is marked 3 while panel (2, 1) is marked 2.

A data set of a choral group of sopranos, altos, tenors, and basses is used to illustrate the multipanel coordinates system. The data includes measurements of the heights of these singers. To compare the distributions of the singers for each of the voice parts we create a histogram panel of the singers in each section. Because

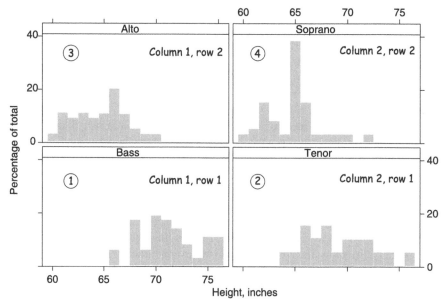

Figure 2.36   Distributions of heights of singers. A multipanel display of histograms where each panel shows the distribution of the heights of the singers in each section of bass, tenor, alto, and soprano

each panel is now associated with a pitch interval, an ordered variable which increases from low to high, we place each panel in a cell of the matrix left-to-right and bottom-to-top. In Fig. 2.36, the panels were ordered by this convention.

Visual Reference Grids   Reference grids—regularly spaced vertical and horizontal gray lines that overlay the graph—were historically used to plot points on a graph or look up their values against horizontal and vertical scales. They tell us nothing about the data. When the graphs are small and being used to summarize data, as they are in multipanel displays, the tick marks are just as effective for look-up. However, in some multipanel displays—to be discussed in later sections—where the panels are alike in all respects, including their vertical and horizontal scales will enhance comparison of patterns.

Grids, when used as common references across identical panels, let us see small movements or changes that would be difficult to detect without them. For example, in Fig. 2.37, graph B is a replicate of graph A, with the exception that a reference grid has been superposed on it. Because the outer frame serves as its own reference point, one might be able to detect that the bar in panel 2 compared with the bar in panel 1 shifted to the right, but it would be hard to see, without the grid, that it has also increased in height.

2.6.1   Histogram Matrix

The *histogram matrix* is a tool for seeing the distribution of many variables in a data set. Because the types and ranges of each variable differ, each panel stands on its

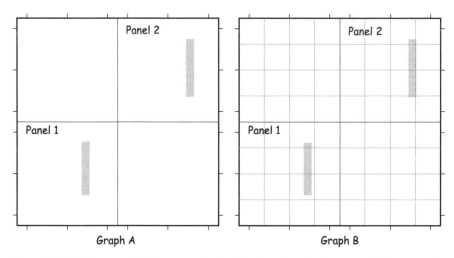

Graph A   Graph B

Figure 2.37   Reference grids. An example showing the value of reference grids in comparing small variations between panels in multipanel displays

own. Reorganizing them has no impact on being able to compare them. Figure 2.38 shows the distributions of measured properties of shell fish.

The histogram matrix can also be used to look at the distributions of one variable over subsets of the data set, where the subsets have been grouped using the classes of a second variable in the set, see Fig. 2.39. The distribution of each subset is displayed in a separate panel. Because groups of the same measurement

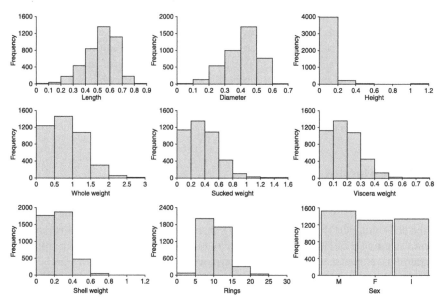

Figure 2.38   Distribution of properties of shell fish. An example showing a histogram matrix of measured properties of shell fish. Source: http://archive.ics.uci.edu/ml/datasets/Abalone

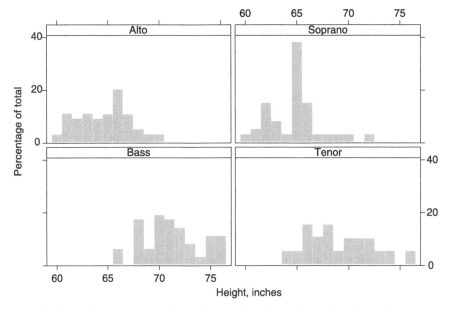

Figure 2.39   A histogram matrix illustrating shared scales and coordinated panels

are being compared, shared scales can be used, allowing us to see the shifts to the left in distribution from bass to alto and tenor to soprano.

That the groups in the bottom row, all men, are taller than the groups in the top row, all women, states the obvious, because histograms are not the best way to compare the distributions of groups. The relationship of height and voice range cannot be clearly seen. However, the point of this example is to illustrate the value of *shared scales* and *coordinated panels* in providing different and useful views of the data.

Only looking at the distribution of the entire set would not have revealed the shift in distribution across groups. The panels provide views of the same variable taken from different angles. Although in this case just one variable was considered, in later sections shared scales and coordinated panels will be discussed, as used in *scatterplot matrices*, *coplots*, and *small multiples*, to help unravel the complexity of data sets with many dimensions.

2.6.2   Scatterplot Matrix

A *scatterplot matrix* efficiently displays scatterplots of pairs of continuous variables in a square matrix by taking advantage of economies of scale. By placing the labels in the cells along the diagonal and moving shared scales to the edges of the matrix as shown in graph B of Fig. 2.40, the scatterplot matrix visually links a row or column of scatterplots. Notice that here are two scatterplots for each unique pair of variables, one on either side of the diagonal. Although this is redundant, it allows you to compare all the pairs that include the variable of interest by looking at just one row or one column without needing to rotate the scale. When reading across, the variable of interest is

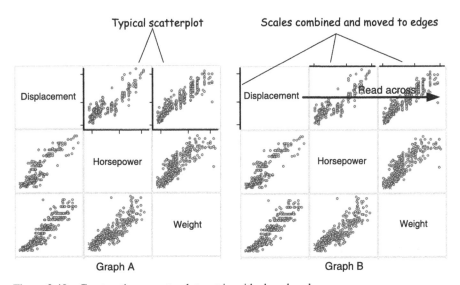

Figure 2.40   Constructing a scatterplot matrix with shared scales

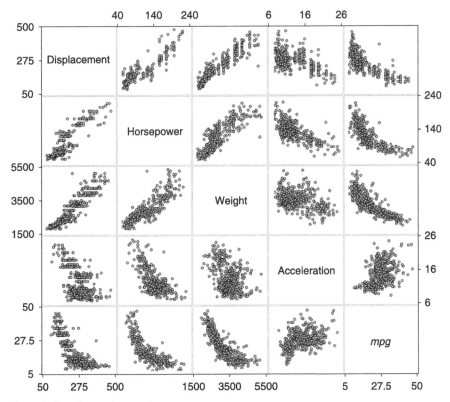

Figure 2.41   Scatterplot matrix

plotted on the vertical axis and the other variables are plotted on the horizontal axis; when reading down, it is plotted on the horizontal axis and the other variables are plotted on the vertical axis.

A complete scatterplot matrix of five variables is shown in Fig. 2.41.

2.6.3 Multiple Box Plot

Multiple box plots allow efficient comparison of distributions for groups of univariate data. The essential five-number summary can be easily compared for each group: center, spread, skewness, tail length, and outlying data points (as discussed in Section 2.4.4). Figure 2.42 shows the box plots for the distributions of groups over the variable miles-per-gallon (*mpg*), where the cars in the data set have been grouped by model year.

2.6.4 Trellis Plot

The *trellis plot*, also known as *small multiple graphs*, is another multipanel display. An example will be used to explain terminology and how it is used. The data set used in this example is a set of 77 breakfast cereal products available in many grocery stores. A simple example is used to illustrate what could be learned through a

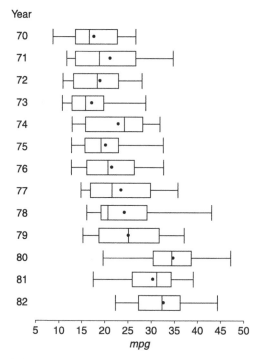

Figure 2.42 A multiple box plot showing a summary of the distribution of groups of cars by year over the variable *mpg*. Source: http://archive.ics.uci.edu/ml/datasets/Auto+MPG

trellis plot about manufacturers' marketing behavior by looking at just three variables from each cold cereal product: the manufacturer, the amount of protein—an indication of the product's nutritional value, and the amount of sugar—an indication of the marketing lure. (Source: http://lib.stat.cmu.edu/DASL/Datafiles/Cereals. html). Figure 2.43 illustrates its use.

In a trellis plot, *panels* are laid out in rows, columns, and pages. This simple example has a single column of three rows. For other data sets, it could be a matrix that flows from page to page. Each panel displays a subset of the data using a display method such as a scatterplot, dot plot, or a multiple box plot. A multiple box plot is used here to show the relationship of two *panel variables*: the manufacturer along the vertical scale and the protein content of the cereal along a shared horizontal scale. The box plot summaries in each panel were conditional on the values of the subset of products selected for that panel.

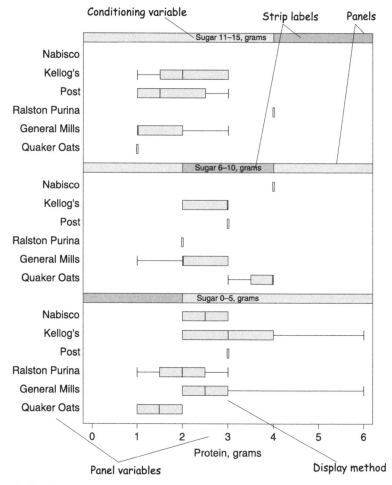

Figure 2.43 An annotated trellis plot

The cereal products were selected for each subset by the *conditioning variable*: the sugar content in the product. In this example just one variable was used, only because using several would have produced many more panels. Each manufacturer's products are divided into three class intervals of low sugar content (0–5 g), medium sugar content (6–10 g), and high sugar content (11–15 g). Each panel focuses on one of these three classes as identified by the *strip label* at the top of the panel. The panels are ordered by the convention of multipanel displays—from the lowest values of the conditioning variable at the bottom to the highest at the top—in order to more easily compare what happens to the panel variables as the sugar content progressively changes. The dark bar inside the strip label of the conditioning variable indicates the values covered by the panel and shows how the variable changes over the trellis.

The ordering of the manufacturers impacts our ability to visually decode the graphic and make comparisons across panels. For example, in Fig. 2.44, the only bar that moved stands out more clearly when comparing the lower and upper panel in the right column, where the bars are ordered by the position of the left edge, than in the left column, where the bars are randomly placed.

The manufacturers are ordered from bottom to top using the median of the subset of each manufacturer's products, starting with the manufacturer with the lowest median. When the medians were the same, the mean was used to break

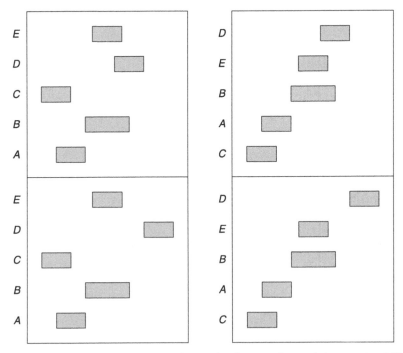

Figure 2.44   An illustration showing the how ordered content in panels impacts the ability to perceive differences between them

the tie. The ordered results quickly reveal some information: Nabisco sells no high sugar content products and its remaining products tend to be relatively high in protein; products higher in protein tend to have less sugar; there are very few products very high in protein with low sugar; a couple of companies, General Mills and Kellog's, have a wide range of products. This is not very profound, but imagine the power a trellis plot could have if it were used to analyze the main and interaction effects of several explanatory variables on response variables using a variety of display methods.

2.7 VISUALIZING GROUPS

2.7.1 Dendrograms

Certain methods for grouping the observations in a data set, such as hierarchical agglomerative clustering, which will be discussed in the next chapter, produce a hierarchy of clusters. A *dendrogram* is a graphic that displays the hierarchy as a tree. Figure 2.45 shows two different ways of drawing a dendrogram. The dendrogram on the right is often used in molecular biology to organize genes and a later section will show how this form can be combined with image maps to construct a cluster image map.

The visual form on the right communicates more than the one on the left. In addition to seeing the groups, you can also see how similar one cluster in the group is to another. If you imagine the vertical line that connects the clusters in a group to be a reference line that extends down to the horizontal scale, you can determine how similar its children clusters are by seeing where the reference line intersects the horizontal scale, see Fig. 2.46. Reference lines closer to the base line are more similar than reference lines farther away. In Fig. 2.46, the group connected at R1, M to N, are more similar to each other than the group connected at R2, H to the cluster A and B.

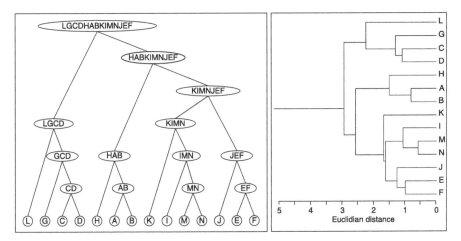

Figure 2.45   Two ways of drawing a dendrogram

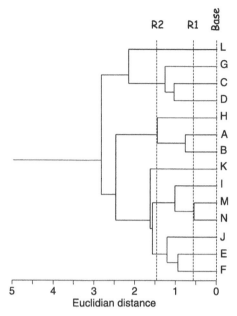

Figure 2.46   An annotated dendrogram

2.7.2   Decision Trees

A *decision tree* is generated through a series of decisions based on certain variables about how to divide a group of observations. The result is a tree, as shown in Fig. 2.47. The process starts with the entire set of observations, or some subset of interest, and uses a scoring function to determine how to split the initial set into two groups. It repeats this process for each of the two resulting groups, and so on until some terminating condition is met. Chapter 5 and Appendix B discusses decision trees.

2.7.3   Cluster Image Maps

A *cluster image map* is a graphic that combines *image maps* with *dendrograms* to display complex, high density data. It is widely used in molecular biology, for example, to look at gene expression patterns in different states of a cell cycle or across different cancer cell lines. Figure 2.48 shows the display of a cluster image map after the data from a microarray of genes has been read and the resulting data set clustered. Each row represents the gene expression levels—a pattern of activity—for a particular gene across multiple experiments.

The *image map*, also referred to as a *heat map*, is a table of colored cells. Each row represents a gene and each column an experiment. The color of each cell, shown in a gray scale in Fig. 2.48, represents the range of values containing the measured value, where the value is mapped onto a spectrum of a color scale that changes from bright green to black to bright red. (To see the figure in color, navigate to the hyperlink http://en.wikipedia.org/wiki/Image:Heatmap.png in your web browser.)

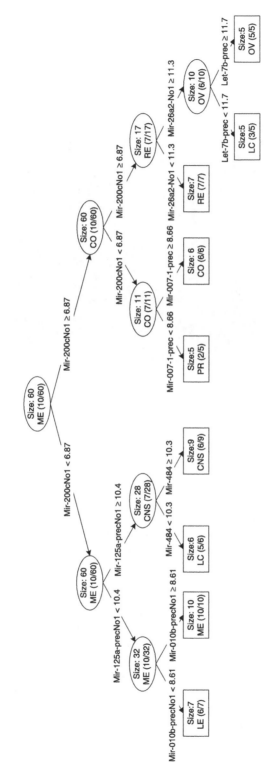

Figure 2.47 Decision tree

61

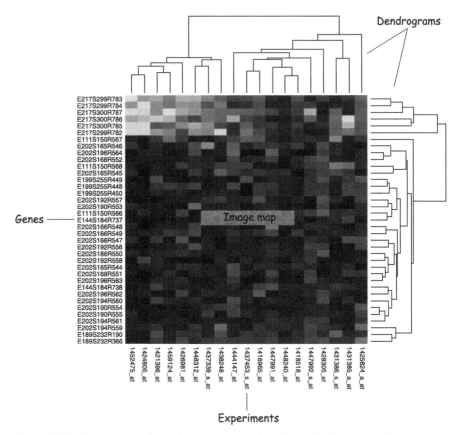

Figure 2.48   An annotated cluster image map showing the results from gene microarray experiments. Original image source: http://en.wikipedia.org/wiki/Image:Heatmap.png

The measured values across each row indicate how much a gene has been expressed relative to some normal expression level within the target cell of the corresponding experiment. In the colored version of Fig. 2.48, the value is mapped to a shade of green if it is lower than expected or to a shade of red if it is higher; black indicates the gene was not expressed at all.

Labels along the left and bottom edges of the image map provide meaningful descriptions of the location of the value in the two-dimensional image. Dendrograms along the right and top edges show the results of clustering the rows and columns, respectively. If a column is a carefully designed experiment that measures, for example, the changes of state in the life of a cell or which genes are expressed within a cancer cell line, then the set of values across a row are a pattern of activity for a gene or, if reading down a column, for a cell state or cancer cell line. Clustering rearranges the rows and columns into groups of similar genes (rows), or states or cancer cell lines (columns) that bring together similar patterns of activity. Because the patterns are color-coded, one can easily see variations within and across the groups that give insight into, for example, the function of genes across cell states or cancer cell lines.

To prepare the data for display in a cluster image map you need to:

- create a data matrix that contains the data values;
- choose a method for determining the distance between a pair of rows or columns in the matrix;
- cluster the rows and cluster the columns;
- rearrange the rows and columns to match the order imposed by their respective hierarchies produced by the clustering.

The distance between a pair of rows or columns for the map above is computed by the *correlation coefficient* distance measure discussed in Section 3.2. The clustering method used to rearrange the data matrix from which the image map is generated is typically generated by the hierarchical agglomerative clustering method introduced in the first book in the series (Myatt, 2007) and discussed in more detail in Section 3.3.

2.8 DYNAMIC TECHNIQUES

2.8.1 Overview

Software that generates good data graphics to describe and summarize a set of numbers eliminates many activities that are mainly clerical or mechanical—drawing graphs, plotting, transforming, and redrawing—and provides a foundation for the analysis and exploration of data. In this context, the computer is used to do what is already being done by hand, but does it much faster. This kind of software is *information software* and its essence is presentation. The important design questions of information software relate to appearance: What information should be presented and how? It requires, in addition to other things, an understanding of graphics design and what has been learned about good design for the print medium. Interaction with information software is primarily navigation. Manipulation software extends information software.

The heart of *manipulation software* is the design of the interaction between the user and the computer. It is the interaction that turns visualization into a computational tool at the center of exploratory data analysis. The interaction can be *active*, where a single action such as the selection of a command from a menu causes a result. Or it can be *dynamic*, where the software responds continuously to interaction with user interface controls. But designing good manipulation software is much harder than designing good information software.

In the physical world, industrial designers shape material into tools—paper clips or hammers, or knives—so that we know how to manipulate them. In the virtual world made possible by computers, software designers of manipulation software—software that lets us use virtual tools to create virtual objects—must know how to make the graphical representation on the display *understandable* (What does this represent to the user?), *available* (how will the user distinguish between tool and objects or determine what can be manipulated, what just informs, and what can do both?), *predictable* (what will happen when the tool is manipulated?), and *comfortable* (how to make it emotionally appealing and easier to learn and use?). Through mouse and keyboard input, the interaction designers must simulate the hand grasping and wielding a tool. Designers must design this simulation, or *direct manipulation*, of virtual tools and objects. In this section, some techniques are discussed, such as brushing, that rely on direct manipulation.

But interaction comes at a cost to the user. For information software, the user has to know what to ask for, how to ask it, how to navigate through complex spaces without getting lost, and remember how to return to important places. For manipulation software they need to know what tools are available, where to find them, and remember in which contexts they work. An important goal of interaction design, therefore, is to reduce interaction. Some ways of doing this are discussed.

2.8.2 Data Brushing

Data brushing, also called *slicing*, is a dynamic interactive technique where the selection by the user of a subset of the data in one data graphic highlights the selected data in that graphic and any others linked to it. Brushing gives new life to histograms which, without interaction, have limited value. The selection of objects, such as points on a scatterplot or bars in a histogram, can be done in a variety of ways, such as by drawing a rubber-band box around the data point to select or selecting an individual object with a mouse click. The highlighting provides feedback that an action has been taken and typically uses color or texture to show the highlighted objects. The panels that respond to highlight in one panel are referred to as being *linked*. Figure 2.49 shows an example of data brushing across linked panels that include histograms and a dendrogram.

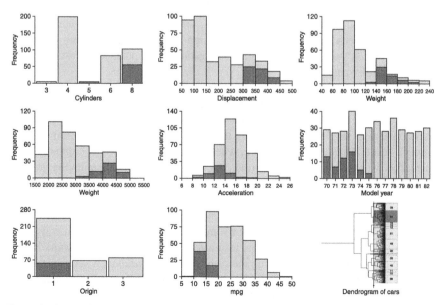

Figure 2.49   Various views of the data set can be linked to allow brushing in one view to be highlighted in other views. The views can be a heterogeneous collection of data graphics

2.8.3   Nearness Selection

The nearness engine proposed by Howard Wainer (Wainer, 2005) allows the user to initiate a search in multivariate data by specifying a point and a distance, or just a distance. For the first case, the selection of a point could be made, for example, on a scatterplot. The result would highlight all points on that scatterplot within the specified distance. If just a distance is provided, the system would identify all groups of points where the points within the group are within the specified distance of each other.

2.8.4   Sorting and Rearranging

As discussed in the section on tables, being able to reorder rows and columns in a table often reveals unexpected relationships. Allowing the user to interact with the data graphic to reorder the data, for example to sort rows by the values in a specific column, is common in spreadsheets and sortable data tables provided by graphical user interface software libraries. Methods include sorting alphabetically, numerically, by date or time, by physical location, and by category or tag.

2.8.5   Searching and Filtering

Searching and filtering are different techniques for narrowing the data to only what is of interest. Searching *includes* only the data that satisfies a query. Filtering *excludes* everything that falls outside the constraints specified by filters. The simplest filters and query techniques give users a choice of which aspects of the data to view. Single-click user interface controls such as checkboxes or radio buttons are interpreted by software as query or filter commands, and the software modifies the graphic in response. Dynamic queries are a logical extension of these simpler controls. User interface controls such as sliders generate a stream of commands to the application software, which continuously adjusts the views in response to these commands.

The best filtering and querying interfaces are highly interactive and iterative, and show the results in context with the surrounding data. Everything in a data graphic becomes a candidate for interaction: Labels of points can be displayed or hidden; legends can reveal more or less explanation on demand; axes, rulers, and scales can show the data for a point or range of values. However, carefully designed graphics can often reduce the need for interaction by answering anticipated questions straightforwardly. Just as with graphics design, simplify and revise; but also be sure to test your ideas with those who will use the tools.

2.9   SUMMARY

Data visualization is a medium for communicating the subtleties and complexities of what is hidden in large sets of measured data. The initial sections of this chapter discussed the general principles of good design: showing the data, simplifying, reducing clutter, revising, and being honest. It showed how ideas borrowed from cognitive psychology, such as the Gestalt principles, and graphics design guide the layout and encoding of data. It described the architecture of a basic graph and the terminology used when referring to its components.

The remaining sections discussed specific graphics and visualization tools for looking at univariate, bivariate, and multivariate data, and groups of data. Tables, various kinds of charts, plots, graphics, and multidisplay panels were described along with examples of their use. Finally, dynamic techniques found in software user interfaces, such as brushing, showed how interaction can extend the reach of visualization for exploratory data analysis.

2.10 FURTHER READING

The following books provide an in-depth analysis of the design and use of data graphics and are essential reading on data visualization: Cleveland (1993, 1994), Tufte (1983), and Hoaglin (2000). Wainer (2005) is an entertaining and serious history of graphical data display from the seventeenth century forward. Wainer (1997) provides insight into the methods and history of presenting data visually. It explains what distinguishes bad design from good and how graphics have been used to distort the truth. Victor (2006) distinguishes between information software (software that helps us learn) and manipulation software (software that helps us create). He argues that most of the software we use is information software and that the long-standing focus on interaction is misguided. Instead, we should focus on the design of context-sensitive information graphics that reduce the need for interaction. We have presented graphs that are commonly used and classified them in large part on the number of dimensions present in the data being displayed. Wilkinson (2005) takes a different approach. He examines the basic elements and structure of graphics, and defines grammatical rules for creating perceivable graphs. His work builds on Bertin (1983).

CHAPTER 3

CLUSTERING

3.1 OVERVIEW

When needing to make sense of a large set of data, the data can be broken down into smaller groups of observations that share something in common. Knowing the contents of these smaller groups helps in understanding the entire data set. Clustering is a widely used and flexible approach to analyzing data, in which observations are automatically organized into groups. Those observations within a particular group are more similar to each other than to observations in other groups. This approach has been successfully utilized in a variety of scientific and commercial applications, including medical diagnosis, insurance underwriting, financial portfolio management, organizing search results, and marketing. For example, clustering has been used by retail organizations to analyze customer data based on historical purchases, along with information about the customer, such as their age or where they live. Customers are grouped using clustering approaches, and specific marketing campaigns are then formulated based on the identified market segments.

A data set of animals will be used to illustrate clustering. Table 3.1 describes a series of animal observations (taken from http://archive.ics.uci.edu/ml/datasets/ Zoo; Murphy and Aha, 1994). Each animal is characterized by a number of variables, including several binary variables, such as whether the animal has hair (*hair*) or produces milk (*milk*). The data set also includes a count of the number of animal legs (*legs*). Cluster analysis organizes the data into groups of similar animals. Using the numeric variables shown in Table 3.1, the data set can be clustered in a number of ways. Figure 3.1 shows one possible grouping of this data. A description of how this particular set was clustered will be discussed later in the chapter. In Fig. 3.1 the animals have been organized into four groups representing the following general categories: (1) mammals, (2) fish and amphibians, (3) invertebrates, and (4) birds. Each group represents a series of similar animals, with the mammals group primarily representing a group of four-legged animals that produce milk, the fish and amphibians group primarily representing a group of aquatic animals, the invertebrates group primarily representing a set of six-legged animals with no backbone, and the birds group primarily representing a set of two-legged airborne animals with feathers.

Based on an initial inspection of the clusters, the grouping may not appear to adequately represent the different types of animals. For example, in Fig. 3.1, an

Making Sense of Data II. By Glenn J. Myatt and Wayne P. Johnson
Copyright © 2009 John Wiley & Sons, Inc.

TABLE 3.1 Portion of a Table Describing Various Animals

Name	Hair	Feathers	Eggs	Milk	Airborne	Aquatic	Predator	Toothed	Backbone	Breathes	Venomous	Fins	Legs	Tail
Aardvark	1	0	0	1	0	0	1	1	1	1	0	0	4	0
Bass	0	0	1	0	0	1	1	1	1	0	0	1	0	1
Boar	1	0	0	1	0	0	1	1	1	1	0	0	4	1
Buffalo	1	0	0	1	0	0	0	1	1	1	0	0	4	1
Carp	0	0	1	0	0	1	0	1	1	0	0	1	0	1
Catfish	0	0	1	0	0	1	1	1	1	0	0	1	0	1
Cheetah	1	0	0	1	0	0	1	1	1	1	0	0	4	1
Chicken	0	1	1	0	1	0	0	0	1	1	0	0	2	1
Clam	0	0	1	0	0	0	1	0	0	0	0	0	0	0

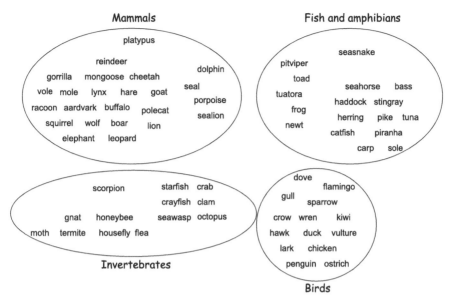

Figure 3.1   Four groups of animals in the data set

octopus is clustered in the same group as a honeybee. To remedy this problem, the set has been organized into 10 groups instead of four, as seen in Fig. 3.2. The number of animals within each of the groups is generally smaller, and the homogeneity among the animals in each group is greater. In Fig. 3.2, the land-living mammals group, the aquatic mammals group, and the platypus are subsets of the previous, larger

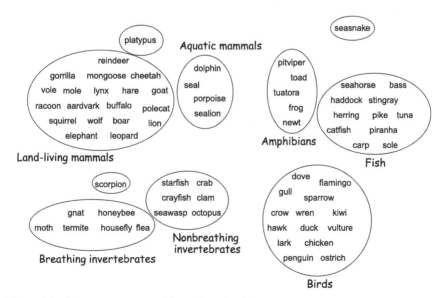

Figure 3.2   Ten groups generated from the animal data set

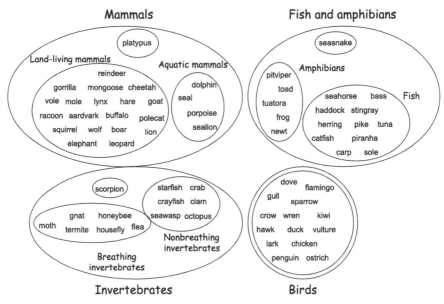

Figure 3.3 The two groupings superimposed

mammals group. The platypus is in a group of its own, since it is both aquatic and lays eggs. Similarly, the fish and amphibians group and the invertebrates group have been divided into smaller groups. The original group of birds remains the same since it already represented a set of similar animals. Figure 3.3 shows the two different groupings superimposed.

Cluster analysis requires neither a prior understanding of the data nor assumptions to be made about it. For example, it is not necessary that the variables follow a normal distribution, nor is it necessary to identify independent and response variables, as is the case when building predictive models. However, care should be taken to select the appropriate variables, including only those relevant to the grouping exercise.

The results from a cluster analysis are groups of observations that share common characteristics. To make cluster analysis useful in many practical situations, these groupings should be analyzed and characterized further. Common approaches to characterizing the clusters include:

- *Summary of cluster*: One approach is to generate a summary for each cluster, such as the mean or mode values for all selected variables. For example, Table 3.2 shows the most common, or mode, values for each variable over the four groups shown in Fig. 3.1. Over all observations in the mammals group, the mode value for *hair* is 1, the mode value for *feathers* is 0, and so on. The table summarizes the contents of each cluster. Alternatively, a representative example from each of the four groups could be used as the summary or a description of the general characteristics of the members. It should be noted that

TABLE 3.2 Mode Values for the Four Groups of Animals

Group	Hair	Feathers	Eggs	Milk	Airborne	Aquatic	Predator	Toothed	Backbone	Breathes	Venomous	Fins	Legs	Tail
Mammals	1	0	0	1	0	0	1	1	1	1	0	0	4	1
Fish and amphibians	0	0	1	0	0	1	1, 0	1	1	1	0	1	0	1
Invertebrates	0	0	1	0	0	0	0	0	0	1	0	0	6	0
Birds	0	1	1	0	1	0	0	0	1	1	0	0	2	1

there could be members of a group that are not well characterized by a representative or a generic class description, such as the clam in the invertebrates groups.

- *Data visualization*: Summarizing the data within a cluster as a series of graphs can also be informative to understanding the contents of the clusters. For example, Fig. 3.4 shows a histogram matrix presenting the variables used for clustering. The outline of the individual histograms represents the entire data set, and the darker shaded regions indicate those observations in the mammals group. Once the graphs are generated, the fact that the mammals cluster generally contains animals with hair, without feathers, that do not lay eggs, which produce milk, and so on, becomes obvious.

- *Naming the clusters*: Manually naming each cluster can be instructive. In the retail example discussed earlier, customer groups could be named according to the characteristics of the interesting groups, such as "Suburban, affluent, and price-sensitive." Naming the clusters is a beneficial exercise because it helps in understanding the results of the clustering and in communicating the results of the clustering to others.

- *Prioritizing value*: In many situations clustering is performed to identify and annotate groupings with certain desired characteristics. Returning to the retail example, historical customer information, such as the products purchased, customer occupations, lifestyle information, and so on, is often clustered for marketing purposes. In this example, a common factor used to assess or prioritize any clusters is the profitability of the customers in the group. Calculating the average profitability for each market segment can be used to direct attention to the most interesting groups.

A useful by-product of cluster analysis is the identification of outliers in the data, or those observations that do not appear similar to others in the data set. These outliers are derived from groups containing a single or a small number of observations. For example, in Fig. 3.2, seasnake, platypus, and scorpion could be

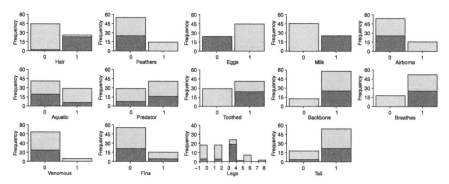

Figure 3.4 Histogram matrix for the animals in the mammals groups compared to the total data set

considered outliers according to the groupings because each belongs to a cluster of one observation. No other animal is considered similar enough to belong in a group with this set at this particular clustering level. In many data mining situations, these unusual observations may be the most interesting. For example, in analyzing customer credit card transactions, these unusual observations may be a result of the use of a stolen credit card. The identification of outliers is also helpful when preparing data, as these observations may be errors warranting further attention.

Another use of clustering is to enable the selection of a smaller number of representatives from the entire data set, a technique known as the selection of *diverse* observations. One approach to selecting 10 diverse observations from the animals grouped in Fig. 3.2 would be to select a single animal from each group, for example: boar, platypus, seal, crayfish, and so on. These examples serve as representatives of all animals in the data set.

In data sets with many variables, clustering can also be used to eliminate highly related variables. Highly related variables are redundant and negatively impact the performance of a predictive model. Grouping the variables and choosing a representative from each group gets rid of the unneeded variables.

Finally, aside from its use in segmenting, clustering is an important tool for learning about the data set during exploration. In addition to identifying groups or outliers, as mentioned above, it helps to see the detail in different contexts. Further, the groups generated by clustering often identify local neighborhoods around which better predictive models can be built (although these models only provide reliable predictions for observations similar to the ones from the model's training set).

A requirement for any clustering exercise is calculating the *distance* between two observations. There are numerous methods for determining this distance. Practical considerations, such as the type of variables (continuous, binary, and so on), dictate the method along with an assessment of how well it matches the problem being addressed. The following chapter outlines different measures for determining these distances.

There are many different types of clustering methods. The selection of a method is influenced by its complexity, the time it takes to generate the clusters, and the number of observations it can process. Certain methods require that a choice be made about the number of clusters to generate prior to the analysis; in other methods this number is determined based on an analysis of the results. Clustering is dependent on the data set being analyzed, and clustering different data sets will invariably result in different groups of observations, even if the data sets contain similar content. In some cases, adding a single new observation to an existing set can have a substantial affect on the groupings. Ties in similarity scores can affect the groupings, and reordering the observations may give different results. There are a number of different types of clustering methods, the most common of which are summarized here:

- *Hierarchical*: These clustering methods organize observations hierarchically. For example, the data set of animals can be organized by four high level

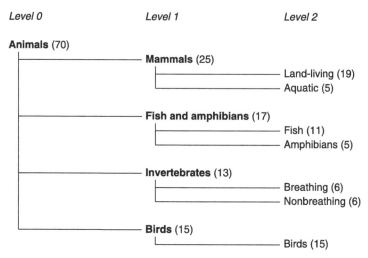

Figure 3.5   Hierarchical relationships for the animal data set

categories which are then subsequently organized into sub-categories, as shown in Fig. 3.3. Figure 3.5 shows how the animal data set can be hierarchically organized into groups and subgroups. The hierarchical organization of groups is flexible since a particular set of groups could be selected at any level. Figures 3.18 and 3.19 illustrate how these two sets of groupings were selected, which will be discussed later in the chapter. After examining the hierarchical relationships, the most useful organization (such as the four groups in level one or the 10 groups in level two) can be chosen from a continuum of levels. Levels at the top partition the data set into fewer, general categories and the levels at the bottom into more, specific categories. The weakness of this method is that it does not scale, which prevents its use in real-time applications and for very large data sets (above 10,000 observations).

- *Partitioned*: These clustering methods assign observations to a single cluster and the number of clusters is set prior to any cluster analysis. For example, Fig. 3.1 could be generated as the result of clustering the animal data set where the number of clusters was preset to four. Although this approach does not provide a flexible hierarchical organization, it is generally faster to compute and can handle larger data sets.

- *Fuzzy*: In many situations, it may not always be easy to assign an observation to a single group because certain observations may share strong associations with several groups. For example, in the animal data set clustered in Fig. 3.1, the platypus is assigned to group 1. Since the platypus lays eggs, the platypus shares traits found in animals in the other groups. Fuzzy clustering, like partitioned-based methods, is based on a predetermined number of clusters; however, it assigns all observations to every cluster. Each observation is then assigned a numeric measure of the degree to which it belongs to each cluster.

Hence, the platypus (as well as all animals), would be assigned a numeric value indicating the degree of association between it and each of the groups.

Different clustering approaches are useful in different situations. It is often helpful to run several methods in order to find an optimal clustering result for the particular problem. Changing the variables, the distance measure, or the clustering parameters can also help create optimal clustering results. The following sections describe the different approaches in greater detail.

3.2   DISTANCE MEASURES

3.2.1   Overview

The starting point for any cluster analysis is the data table or data matrix. For example, suppose a retail company has gathered information about its customers, such as addresses, income, occupations, and so on, along with information concerning the categories of purchases made by those customers throughout the year. Table 3.3 illustrates five customer entries.

Because cluster analysis can only operate on numerical data, any nonnumeric values in the data tables should be converted to numbers. Table 3.3 can be easily transformed into numeric values, as shown in Table 3.4. In this example, categorical fields with text entries were converted into dummy variables with values of zero or one, indicating the presence or absence of a specific category. The *geographic* variable was converted to the variable *urban*, the variable *employer* was converted to the variables *technology*, *services*, and *legal*, and the variable *gender* was converted to the variable *female*.

It is also common to normalize the data to standard ranges, such as between zero and one or using a z-score (the number of standard deviations above or below the mean), to ensure a variable is not considered more important as a consequence of the range on which it is measured. In Table 3.5 the variables *income*, *age*, *children*, *electronics*, *groceries*, and *clothes* have all been normalized such that all variables are in the range $0-1$.

After deciding which variables to use, a method for determining the distance must be selected. There are many methods to choose from such as the *euclidean* distance, the *Manhatten* distance, and so on. The choice is based on content, that is, which method appears to be the most appropriate to the observations being analyzed within the particular problem. Sections $3.2.2-3.2.4$ describe different distance measures that are often used.

Distance measures generally share certain properties. First, any distance measure between two observations is greater than or equal to zero. Second, the distance between observations A and B is the same as the distance between observations B and A. Third, if the distance is zero, there is no difference between the two observations, that is, the two observations have the same values for all variables. Finally, the distance between A and B ($d_{A,B}$) satisfies the following assumptions with respect to a third observation (C), based on the distance between A and C ($d_{A,C}$) and the

TABLE 3.3 Information on Customers

Customer	Geographic	Income	Employer information	Age	Children	Gender	Electronics	Groceries	Clothes
A	Urban	85,000	Technology	31	0	Male	15	1	0
B	Suburbs	38,000	Services	43	3	Female	1	33	8
C	Suburbs	150,000	Legal	51	2	Female	5	0	0
D	Suburbs	32,000	Services	49	1	Female	2	29	9
E	Urban	104,000	Technology	36	0	Male	8	0	1

TABLE 3.4 Information on Customers Converted to Numbers

Customer	Urban	Income	Technology	Services	Legal	Age	Children	Female	Electronics	Groceries	Clothes
A	1	85,000	1	0	0	31	0	0	15	1	0
B	0	38,000	0	1	0	43	3	1	1	33	8
C	0	150,000	0	0	1	51	2	1	5	0	0
D	0	32,000	0	1	0	49	1	1	2	29	9
E	1	104,000	1	0	0	36	0	0	8	0	1

TABLE 3.5 Information on Customers, Normalized to the Range 0 to 1

Customer	Urban	Income	Technology	Services	Legal	Age	Children	Female	Electronics	Groceries	Clothes
A	1	0.449	1	0	0	0	0	0	1	0.03	0
B	0	0.051	0	1	0	0.6	1	1	0	1	0.89
C	0	1.000	0	0	1	1	0.67	1	0.29	0	0
D	0	0.000	0	1	0	0.9	0.33	1	0.071	0.88	1
E	1	0.610	1	0	0	0.25	0	0	0.5	0	0.11

distance between B and C $(d_{B,C})$:

$$d_{A,B} \leq d_{A,C} + d_{B,C}$$

Clustering methods generally require a *distance matrix* or *dissimilarity matrix* (D) as input. This matrix is a square n-by-n matrix, where n is the number of observations in the data set to be clustered. This distance matrix has the following format:

$$D = \begin{bmatrix} 0 & d_{1,2} & d_{1,3} & \cdots & d_{1,n} \\ d_{2,1} & 0 & d_{2,3} & \cdots & d_{2,n} \\ d_{3,1} & d_{3,2} & 0 & \cdots & d_{3,n} \\ \cdots & \cdots & \cdots & \cdots & \cdots \\ d_{n,1} & d_{n,2} & d_{n,3} & \cdots & 0 \end{bmatrix}$$

where $d_{i,j}$ is the distance value between pairs of observations; for example, $d_{1,2}$ is the distance between the first and second observation. The diagonal values are all zero distances since there is no distance between two identical observations. The distance matrix for the observations in Table 3.5, using the *Gower* distance (described in Section 3.3.4) over all variables, is:

$$D = \begin{bmatrix} & A & B & C & D & E \\ A & 0 & 0.865 & 0.754 & 0.836 & 0.178 \\ B & 0.865 & 0 & 0.677 & 0.227 & 0.814 \\ C & 0.754 & 0.677 & 0 & 0.67 & 0.688 \\ D & 0.836 & 0.227 & 0.67 & 0 & 0.777 \\ E & 0.178 & 0.814 & 0.688 & 0.777 & 0 \end{bmatrix}$$

The smallest distance is between A and E (0.178), which makes sense because both observations are urban-living males working in technology who primarily purchase electronic products. The largest distance is between A and B (0.865), which also makes sense because these observations live in different areas, have different income levels, work in different industries (A in technology, B in services), and so on.

3.2.2 Numeric Distance Measures

As mentioned above, a table of all numeric data can be clustered using a variety of distance measures. Table 3.6 will be used as an example in this section and contains

TABLE 3.6   Six Numeric Variables Describing Five Customers

Customer	Income	Age	Children	Electronics	Groceries	Clothes
A	0.449	0	0	1	0.03	0
B	0.051	0.6	1	0	1	0.89
C	1.000	1	0.67	0.29	0	0
D	0.000	0.9	0.33	0.071	0.88	1
E	0.610	0.25	0	0.5	0	0.11

a subset of variables over the same five observations as Table 3.5. This table is limited to six normalized variables (*income, age, children, electronic, groceries,* and *clothes*).

Distance measures compute a number for any pair of observations. These measures compare the values for each of the variables in the two observations and compute a distance based on some function relating to the differences between these values. The following describes a number of distance measures.

Euclidean The euclidean distance is one of the most common distance functions. If a data set only has two variables, then the euclidean distance would calculate the physical distance between the two points plotted on a scatterplot. The following formula calculates a euclidean distance between two observations (*p* and *q*), measured over *n* variables:

$$d_{p,q} = \sum_{i=1}^{n} \sqrt{(p_i - q_i)^2}$$

To demonstrate how this formula works, the euclidean distance between observations *A* and *B* using the data in Table 3.6 is calculated as follows:

$$d_{A,B} = \sqrt{\begin{array}{l}(0.449 - 0.051)^2 + (0 - 0.6)^2 + (0 - 1)^2 \\ + (1 - 0)^2 + (0.03 - 1)^2 + (0 - 0.89)^2\end{array}}$$

$$d_{A,B} = 2.061$$

Square euclidean The square euclidean is the sum of the squares of the difference between the two observations, and it is calculated using the following formula:

$$d_{p,q} = \sum_{i=1}^{n} (p_i - q_i)^2$$

To calculate the square euclidean distance between observations *A* and *B* using the data in Table 3.6, the following calculation is made:

$$d_{A,B} = (0.449 - 0.051)^2 + (0 - 0.6)^2 + (0 - 1)^2 + (1 - 0)^2$$
$$+ (0.03 - 1)^2 + (0 - 0.89)^2$$
$$d_{A,B} = 4.249$$

Manhattan The Manhattan distance, which is also called the *city block* distance, is the sum of the absolute distances between the variables, which is always a positive value representing the difference. The formula is:

$$d_{p,q} = \sum_{i=1}^{n} |p_i - q_i|$$

Calculating the manhattan distance between observations A and B using the data in Table 3.6, is done as follows:

$$d_{A,B} = |0.449 - 0.051| + |0 - 0.6| + |0 - 1| + |1 - 0|$$
$$+ |0.03 - 1| + |0 - 0.89|$$
$$d_{A,B} = 4.856$$

Maximum To calculate the maximum distance between two observations, the absolute difference between each variable is determined and the highest difference is selected:

$$d_{p,q} = \max |p_i - q_i|$$

The maximum distance between observations A and B, using the data in Table 3.6, is 1.0, because that is the difference between the values in the variable *children* (0 minus 1), as well as *electronics* (1 minus 0).

Minkowski The Minkowski distance is a general distance formula, where the order, or λ, can take any positive value. When λ is ∞, the Minkowski distance is the same as the maximum distance. When λ is 1 the Minkowski distance is the same as the Manhattan distance. And when λ is 2 the Minkowski distance is the same as the euclidean distance. The following formula is used to calculate the Minkowski distance:

$$d_{p,q} = \sqrt[\lambda]{\sum_{i=1}^{n} |p_i - q_i|^{\lambda}}$$

Using the data in Table 3.6, the following example demonstrates how to calculate the Minkowski distance between A and B, where λ is equal to 3:

$$d_{A,B} = \sqrt[3]{\frac{(0.449 - 0.051)^3 + (0 - 0.6)^3 + (0 - 1)^3 + (1 - 0)^3}{+(0.03 - 1)^3 + (0 - 0.89)^3}}$$
$$d_{A,B} = 1.573$$

While the above examples only compared observations A and B, Table 3.7 compares the distances over all observations in Table 3.6 based on the distance measures outlined in this section. Each distance measure calculates different values, and in some cases yields different results. For example, the fifth highest euclidean distance is between C and D, yet the fifth highest Manhattan distance is between D and E.

3.2.3 Binary Distance Measures

A number of measures have been developed to calculate distances when the observations being compared use all binary variables; that is, variables with values of

TABLE 3.7 Distances between all Pairs of Observations (from Table 3.6) Across Seven Distance Measures

	Euclidean	Square euclidean	Manhatten	Maximum	Minkowski $(\lambda = 3)$	Minkowski $(\lambda = 4)$	Minkowski $(\lambda = 5)$
A,B	2.061	4.249	4.856	1	1.573	1.383	1.284
A,C	1.502	2.258	2.962	1	1.222	1.115	1.064
A,D	1.924	3.705	4.459	1	1.484	1.312	1.223
A,E	0.593	0.351	1.052	0.5	0.526	0.509	0.503
B,C	1.744	3.043	3.857	1	1.389	1.254	1.185
B,D	0.754	0.569	1.321	0.666	0.688	0.673	0.669
B,E	1.813	3.29	4.187	1	1.411	1.262	1.188
C,D	1.714	2.939	3.526	1	1.397	1.271	1.203
C,E	1.103	1.217	2.131	0.75	0.923	0.856	0.823
D,E	1.628	2.651	3.789	0.888	1.259	1.121	1.053

either zero or one. Using binary variables enables distances to be calculated for categorical values because they can be easily converted to binary dummy variables. For each variable, the values in the two observations are compared to determine whether they are the same or different. Table 3.8 illustrates the four possible ways the values between two observations (p_i, q_i) can be compared.

Similarity measures how alike two observations are to each other, with high similarity values representing situations when the two observations are alike. This contrasts with distance (or dissimilarity) measures, where low values indicate the observations are alike. The similarity and distance calculations for binary variables are based on the number of common and different values in the four situations summarized in Table 3.8 for two observations p and q, over all variables. The following summarizes these counts:

- a: the number of variables where the value for both p and q is one (A in Table 3.8);
- b: the number of variables where the value for p is one and the value for q is zero (B in Table 3.8);
- c: the number of variables where the value for p is zero and the value for q is one (C in Table 3.8);
- d: the number of variables where the value for both p and q is zero (D in Table 3.8).

TABLE 3.8 Four Alternatives (A, B, C, and D) for Comparing Two Binary Values

		q_i	
		1	0
p_i	1	A	B
	0	C	D

TABLE 3.9   Five Customers with Data for Five Binary Variables

Customer	Urban	Technology	Services	Legal	Female
A	1	1	0	0	0
B	0	0	1	0	1
C	0	0	0	1	1
D	0	0	1	0	1
E	1	1	0	0	0

Table 3.9 will be used to illustrate binary distance measures, and this table shows the same five observations taken from Table 3.5 over the following five binary variables: *urban, technology, services, legal,* and *female*.

Using Table 3.9, the values for a, b, c, and d for customers B and C are calculated as:

$a = 1$ (since *female* is both one in observations B and C);

$b = 1$ (since *services* is one in observation B and zero in observation C);

$c = 1$ (since *legal* is zero in observation B and one in observation C);

$d = 2$ (since *urban* and *technology* are both zero in observations B and C).

Since these four counts cover all possible situations, summing a, b, c, and d should equal the total number of variables selected, which is five in this example.

The following section explains a selection of similarity and corresponding distance measures that operate on binary variables. Both similarity and distance measures are presented for completeness.

Simple matching   This method calculates the number of common ones or zeros as a proportion of all variables. The following formula is used to calculate the similarity coefficients ($s_{p,q}$):

$$s_{p,q} = \frac{a + d}{a + b + c + d}$$

The corresponding distance calculation is:

$$d_{p,q} = 1 - \frac{a + d}{a + b + c + d}$$

Using Table 3.9, the simple distance between B and C is:

$$d_{B,C} = 1 - \frac{1 + 2}{1 + 1 + 1 + 2}$$

$$d_{B,C} = 0.4$$

Jaccard   This method calculates the proportion of common ones against the total number of values that are one in either or both observations. Or put another way, the method does not incorporate d (the number of values where the variables are both zero). The following formula is used to calculate the similarity coefficients ($s_{p,q}$):

$$s_{p,q} = \frac{a}{a+b+c}$$

The corresponding distance calculation is:

$$d_{p,q} = \frac{b+c}{a+b+c}$$

Using Table 3.9, the Jaccard distance between B and C is:

$$d_{B,C} = \frac{1+1}{1+1+1}$$

$$d_{B,C} = 0.67$$

Russell and Rao   This method calculates the proportion of variables where one is common in both observations against the total number of variables. Whereas the Jaccard method disregarded the count for where both values are zero (d), the Russell and Rao method incorporates it into the calculation. The following formula is used to calculate the similarity coefficients ($s_{p,q}$):

$$s_{p,q} = \frac{a}{a+b+c+d}$$

The corresponding distance calculation is:

$$d_{p,q} = 1 - \frac{a}{a+b+c+d}$$

Using Table 3.9, the Russel and Rao distance between B and C is:

$$d_{B,C} = 1 - \frac{1}{1+1+1+2}$$

$$d_{B,C} = 0.8$$

Dice   The *Dice* method takes into account the number of variables in the two observations which have one in common and weighs this against the number of variables with one in either or both observations. This method is similar to Jaccard; however, the Dice formula puts greater emphasis on the common ones (a). The following formula is used to calculate the similarity coefficients ($s_{p,q}$):

$$s_{p,q} = \frac{2a}{2a+b+c}$$

The corresponding distance calculation is:

$$d_{p,q} = \frac{b+c}{2a+b+c}$$

Using Table 3.9, the Dice distance between customer B and C is:

$$d_{B,C} = \frac{1+1}{2 \times 1 + 1 + 1}$$

$$d_{B,C} = 0.5$$

Rogers and Tanimoto   Unlike the methods described above, this method takes into account when both variables have common zeros or ones. The following formula is used to calculate the similarity coefficients ($s_{p,q}$):

$$s_{p,q} = \frac{a+d}{a + 2(b+c) + d}$$

The corresponding distance calculation is:

$$d_{p,q} = \frac{2(b+c)}{a + 2(b+c) + d}$$

Using Table 3.9, the Rogers and Tanimoto distance between B and C is:

$$d_{B,C} = \frac{2(1+1)}{1 + 2(1+1) + 2}$$

$$d_{B,C} = 0.57$$

Table 3.10 compares the distance scores for all pairs of observations from Table 3.9 calculated using the simple, Jaccard, Russell and Rao, Dice, and Rogers and Tanimoto methods. The specific distance measure should be selected that most closely represents the distance between the observations for the specific problem being addressed.

Other binary similarity and distance measures can be seen in Table 3.11.

TABLE 3.10   Comparison of Pairs of Distances for Five Customers (from Table 3.9) Over Five Binary Distance Measures

	Simple	Jaccard	Russell and Rao	Dice	Rogers and Tanimoto
A,B	0.8	1	1	1	0.89
A,C	0.8	1	1	1	0.89
A,D	0.8	1	1	1	0.89
A,E	0	0	0.6	0	0
B,C	0.4	0.67	0.8	0.5	0.57
B,D	0	0	0.6	0	0
B,E	0.8	1	1	1	0.89
C,D	0.4	0.67	1	0.5	0.57
C,E	0.8	1	1	1	0.89
D,E	0.8	1	1	1	0.89

TABLE 3.11 Additional Binary Similarity and Distance Measures

Name	Similarity	Distance
Pearson	$\dfrac{ad - bc}{\sqrt{(a+b)(a+c)(d+b)(d+c)}}$	$\dfrac{1}{2} - \dfrac{ad - bc}{2\sqrt{(a+b)(a+c)(d+b)(d+c)}}$
Yule	$\dfrac{ad - bc}{ad + bc}$	$\dfrac{bc}{ad + bc}$
Sokal–Michener	$\dfrac{a+d}{a+b+c+d}$	$\dfrac{2b + 2c}{a + 2b + 2c + d}$
Kulzinsky	$\dfrac{a}{b+c}$	$\dfrac{2b + 2c + d}{a + 2b + 2c + d}$

3.2.4 Mixed Variables

Since data sets rarely contain solely continuous variables or solely discrete variables, additional methods are needed to cluster data sets that include this mixture of variables. This section describes the *Gower* distance measure used for this case.

Gower   This measure is calculated with the following formula for two observations p and q, over i variables:

$$d_{p,q} = \sqrt{\frac{\sum\limits_{i=1}^{n} w_i d_i^2}{\sum\limits_{i=1}^{n} w_i}}$$

where w_i is a weight for the ith variable, and it takes the value one when both values are known; otherwise it is zero. d_i^2 is the square of the distance between the ith value of the two observations (p_i and q_i), where:

$$d_i = \frac{|p_i - q_i|}{R_i}$$

and R_i is the range over all values of the ith variable. For categorical variables, d_i is zero if the p_i and q_i are the same, otherwise it is 1.

To illustrate the calculation of the Gower distance, the data in Table 3.12 will be used, and observations B and C will be compared. Since there are no missing values in any of the observations, the weights (w_i) will all be 1. Table 3.13 shows the intermediate calculations necessary to calculate the Gower distance.

Using Table 3.13 $\sum_{i=1}^{n} w_i$ is determined to be 11, and $\sum_{i=1}^{n} w_i d_i^2$ to be 5.04. The Gower distance is then calculated as:

$$d_{B,C} = \sqrt{\frac{\sum\limits_{i=1}^{n} w_i d_i^2}{\sum\limits_{i=1}^{n} w_i}}$$

$$d_{B,C} = \sqrt{\frac{5.04}{11}} = 0.68$$

TABLE 3.12 Five Customers Described Using a Mixture of Discrete and Continuous Variables

Customer	Urban	Income	Technology	Services	Legal	Age	Children	Female	Electronics	Groceries	Clothes
A	1	0.449	1	0	0	0	0	0	1	0.03	0
B	0	0.051	0	1	0	0.6	1	1	0	1	0.89
C	0	1.000	0	0	1	1	0.67	1	0.29	0	0
D	0	0.000	0	1	0	0.9	0.33	1	0.071	0.88	1
E	1	0.610	1	0	0	0.25	0	0	0.5	0	0.11

TABLE 3.13 Intermediate Calculations for the Gower Distance

	B	C	$d = \dfrac{\lvert p_i - q_i \rvert}{R_i}$	W	$w_i d_i^2$
Urban	0	0	0	1	0
Income	0.051	1	0.949	1	0.9
Technology	0	0	0	1	0
Services	1	0	1	1	1
Legal	0	1	1	1	1
Age	0.6	1	0.4	1	0.16
Children	1	0.67	0.33	1	0.11
Female	1	1	0	1	0
Electronics	0	0.29	0.29	1	0.08
Groceries	1	0	1	1	1
Clothes	0.89	0	0.89	1	0.79
				11	5.04

3.2.5 Other Measures

There are many other measures that can be incorporated into clustering approaches. The following section describes four additional methods: the use of the Mahalanobis, correlation coefficients, cosine, and Canberra measures.

Mahalanobis The Mahalanobis distance takes into account correlations within a data set between the variables. Unlike most other distance measures, this method is not dependent upon the scale on which the variables are measured. The following formula is used to calculate the Mahalanobis distance:

$$d_{p,q} = \sqrt{(p - q)S^{-1}(p - q)^{\mathsf{T}}}$$

The superscript "-1" represents the inverse matrix and "T" is the transformed matrix, described in Appendix A. S is the covariance matrix and is calculated based on all pairs of variables using the following equation:

$$S_{j,k} = \frac{1}{n-1} \sum_{i=1}^{n} (X_{ij} - \bar{X}_j)(X_{ik} - \bar{X}_k)$$

Under this formula $S_{j,k}$ are the values in the covariance matrix, X is the original data table with X_{ij} as the value at column j and row i, n is the number of observations, $\bar{X}_j$ is the average for column j, and $\bar{X}_k$ is the average for row k.

Correlation coefficient Using clustering it is also possible to group variables based on the linear relationship between pairs of variables (r). Using the following formula, variables x and y can be compared:

$$r = \frac{\sum_{i=1}^{n} (x_i - \bar{x})(y_i - \bar{y})}{(n-1)s_x s_y}$$

Under this formula, $\bar{x}$ is the mean of the x variable and $\bar{y}$ is the mean of the y variable. The number of observations is n. Additionally, s_x is the standard deviation for x and s_y is the standard deviation for y.

Cosine   Using the cosine formula is often effective when clustering vectors of features, such as terms or words that relate to documents, which are often sparse. The formula for cosine (cs) where p and q represent the vectors is:

$$cs = \frac{\sum_{i=1}^{n}(p_i q_i)}{\sqrt{\sum_{i=1}^{n} p_i^2 \sum_{i=1}^{n} q_i^2}}$$

Canberra   The Canberra metric is the sum of the fractional differences for each variable. This method is sensitive to changes close to zero, and it is calculated using the following formula:

$$d_{p,q} = \sum_{i=1}^{n} \frac{|p_i - q_i|}{(|p_i| + |q_i|)}$$

3.3   AGGLOMERATIVE HIERARCHICAL CLUSTERING

3.3.1   Overview

The agglomerative hierarchical approach to clustering starts with each observation as its own cluster and then continually groups the observations into increasingly larger groups. This results in a hierarchical organization of the data which can be used to characterize the data set, and from this hierarchy specific groups can be selected. This method is useful and a widely used approach to analyzing data; however, the analysis can be time-consuming, and it is usually limited to smaller data sets because the method requires the complete distance matrix to be calculated.

Like all clustering approaches, the first step in performing agglomerative hierarchical clustering is to select the variables and observations to cluster. For reasons discussed earlier, the data should also be normalized. Table 3.14 shows five observations: A, B, C, D, and E. Each observation is measured over

TABLE 3.14   Five Observations Measured Over Five Variables

	Percentage body fat	Weight	Height	Chest	Abdomen
A	12.3	154.25	67.75	93.1	85.2
B	31.6	217	70	113.3	111.2
C	22.2	177.75	68.5	102	95
D	14.1	176	73	96.7	86.5
E	23.6	197	73.25	103.6	99.8

TABLE 3.15   Normalized Data

	Percentage body fat	Weight	Height	Chest	Abdomen
A	0.0	0.0	0.0	0.0	0.0
B	1.0	1.0	0.409	1.0	1.0
C	0.513	0.375	0.136	0.441	0.377
D	0.00932	0.347	0.955	0.178	0.050
E	0.585	0.681	1.0	0.520	0.562

five variables: *percentage body fat*, *weight*, *height*, *chest*, and *abdomen*. In Table 3.15, the data is normalized to the range 0–1, using the min–max normalization, discussed in Chapter 1.

In the first stage, each observation is treated as a single cluster. The next step is to compute a distance between each of these single-observation clusters. Section 3.2 outlined a series of methods for computing distances between pairs of observations. Table 3.16 illustrates a distance matrix where the distance between each combination of single-observation clusters is recorded. In this example, the euclidean distance was selected to calculate the distance between observations.

The process of agglomerative hierarchical clustering starts with these single-observation clusters and progressively combines pairs of clusters, forming smaller numbers of clusters that contain more observations. The process of combining clusters is repeated until there is a single cluster containing all observations in the data set. There are a number of alternative approaches to joining clusters together, such as single linkage, complete linkage, average linkage, and so on. These approaches are described in the following sections.

3.3.2   Single Linkage

The single linkage agglomerative hierarchical clustering method joins pairs of clusters together based on the shortest distance between two groups, as illustrated in Fig. 3.6. The single linkage method allows clusters to be easily joined; however, it may prevent the detection of clusters that are not easily separated. The single linkage method is best used in situations where the clusters are long and tubular in shape, since an observation will become part of a cluster with just a single nearest neighbor. In practice, the single linkage approach is the least frequently used joining method.

TABLE 3.16   Distance Matrix Using the Euclidean Distance

	{A}	{B}	{C}	{D}	{E}
{A}	0	2.041	0.871	1.036	1.547
{B}	2.041	0	1.185	1.768	1.022
{C}	0.871	1.185	0	1.011	0.941
{D}	1.036	1.768	1.011	0	0.857
{E}	1.547	1.022	0.941	0.857	0

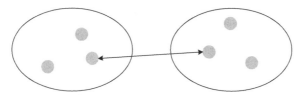

Figure 3.6   Using the shortest distance to decide whether to join the two groups

The data from Table 3.16 will be used to illustrate the single linkage approach. Because all agglomerative methods start with each observation as an individual cluster, the first step is to identify a pair of clusters to combine. Since the distance between $\{D\}$ and $\{E\}$ is the smallest, these single observation clusters are the first to be combined into a single group $\{D,E\}$. A new distance matrix is calculated, as shown in Table 3.17, for the four remaining clusters: $\{A\}$, $\{B\}$, $\{C\}$, and $\{D,E\}$. In determining the distance between the new cluster $\{D,E\}$ and the other clusters, the shortest distance is selected. For example, the distance between $\{B\}$ and $\{D\}$ is 1.768, and the distance between $\{B\}$ and $\{E\}$ is 1.022, and hence the distance between $\{D,E\}$ and $\{B\}$ is the smallest distance, or 1.022.

Figure 3.7 illustrates the first step in the construction of a clustering dendrogram. Since it was determined that $\{D\}$ and $\{E\}$ should be initially combined, the

TABLE 3.17   Distance Matrix for the Four Groups Using the Single Linkage Criteria

	$\{A\}$	$\{B\}$	$\{C\}$	$\{D,E\}$
$\{A\}$	0	2.041	0.871	1.036
$\{B\}$	2.041	0	1.185	1.022
$\{C\}$	0.871	1.185	0	0.941
$\{D,E\}$	1.036	1.022	0.941	0

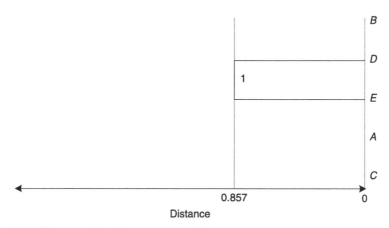

Figure 3.7   First step of dendrogram construction

TABLE 3.18 Distance Matrix for the Three Groups Using the Single Linkage Criteria

	{B}	{D,E}	{A,C}
{B}	0	1.022	1.185
{D,E}	1.022	0	0.941
{A,C}	1.185	0.941	0

two observations are joined at 1 in Fig. 3.7. The horizontal axes represent the distances at which the joining takes place. The vertical line joining the two observations is at a distance of 0.857.

The next step in the process is to identify the next group to combine. The smallest distance in Table 3.17 is 0.871, and hence the groups {A} and {C} are combined. The groups containing the individual observations A and C are no longer separately considered. A new distance matrix is computed containing distances between the three remaining groups: {B}, {D,E}, and {A,C}. This distance matrix is shown in Table 3.18.

Figure 3.8 illustrates the second step in the construction of a clustering dendrogram. Since it was determined that {A} and {C} should be next combined, the two observations are joined at 2 in the figure. The horizontal axes represent the distances at which the joining takes place. The vertical line joining the two observations is at a distance of 0.871.

Since the distance between {A,C} and {D,E} is the smallest, these two groups are combined. Table 3.19 shows the distance matrix for the remaining two groups, {B} and {A,C,D,E}.

Figure 3.9 illustrates the next step in the construction of a clustering dendrogram. Since it was determined that the {A,C} and the {D,E} groups should be next combined, the two groups are joined at 3 in the figure. The horizontal axes represent the distances at which the joining takes place. The vertical line joining the two groups is at a distance of 0.941.

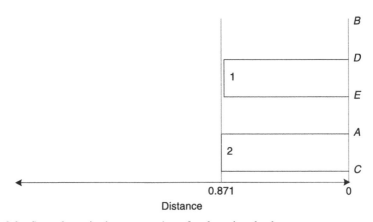

Figure 3.8   Second step in the construction of a clustering dendrogram

TABLE 3.19   Distance Matrix for the Two Groups Using the Single Linkage Criteria

	{B}	{A,C,D,E}
{B}	0	1.022
{A,C,D,E}	1.022	0

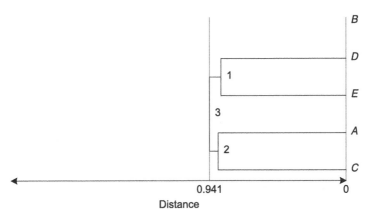

Figure 3.9   Third step in the construction of a clustering dendrogram

Groups {B} and {A,C,D,E} are combined in the last step at a distance of 1.022, based on Table 3.19.

Figure 3.10 illustrates the final step in the construction of a clustering dendrogram. The remaining groups are combined: {B} and {A,B,C,D} at 4 on the figure. The horizontal axes represent the distances at which the joining takes place. The vertical line joining the two groups is at a distance of 1.022.

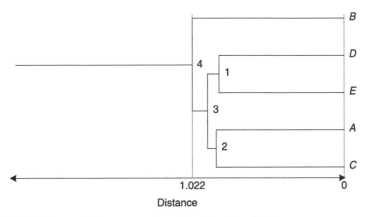

Figure 3.10   Final step in the construction of the clustering dendrogram

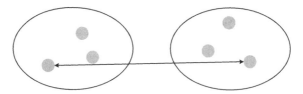

Figure 3.11   Using the farthest distance to decide whether to join the two groups

3.3.3   Complete Linkage

When performing an agglomerative hierarchical clustering using the complete linkage method, the distance of the two observations that are the furthest away is used to determine a distance between two clusters. For example, in Fig. 3.11, two clusters of observations are evaluated and the distance between the furthest two observations is used to establish a distance between the two clusters. This approach often results in a smaller number of clusters and can prematurely divide larger clusters. Under this methodology, clusters are often compact in nature, and the maximum within cluster distance (or cluster diameter) is known.

The first step in this cluster analysis is similar to the single linkage method. Using the original distance matrix from Table 3.16, the first new group $\{D,E\}$ is formed since the 0.857 distance between $\{D\}$ and $\{E\}$ is the smallest. A new distance matrix is now calculated, as shown in Table 3.20, using the complete linkage rule. The distance between $\{A\}$ and $\{D\}$ (1.036) and the distance between $\{A\}$ and $\{E\}$ (1.547) will be used for illustration. The distance between $\{A\}$ and $\{D,E\}$ is determined using the complete linkage joining rule which is the farthest distance, that is 1.547. Groups $\{D\}$ and $\{E\}$ are no longer considered separately since they have been combined into the new group $\{D,E\}$.

Table 3.21 shows the next step where a new group is formed: $\{A,C\}$. Distances are computed using the complete linkage rules. Only three groups are now considered: $\{B\}$, $\{D,E\}$, and $\{A,C\}$.

Table 3.22 presents a distance matrix for the two remaining groups: $\{B\}$ and $\{A,C,D,E\}$. In the final step the two groups are combined at a distance of 2.041.

Figure 3.12 illustrates a dendrogram showing the hierarchical relationships between the groups. The structure of the dendrogram is different from Fig. 3.10 since the distances at which the clusters are merged are different as a result of using the complete linkage joining rules instead of the single linkage method. $\{D\}$

TABLE 3.20   Distance Matrix for the Four Groups Using the Complete Linkage Criteria

	$\{A\}$	$\{B\}$	$\{C\}$	$\{D,E\}$
$\{A\}$	0	2.041	0.871	1.546
$\{B\}$	2.041	0	1.185	1.767
$\{C\}$	0.871	1.185	0	1.01
$\{D,E\}$	1.546	1.767	1.01	0

TABLE 3.21   Distance Matrix for the Three Groups Using the Complete Linkage Criteria

	{B}	{D,E}	{A,C}
{B}	0	1.767	2.041
{D,E}	1.767	0	1.546
{A,C}	2.041	1.546	0

TABLE 3.22   Distance Matrix for the Two Groups Using the Complete Linkage Criteria

	{B}	{A,C,D,E}
{B}	0	2.041
{A,C,D,E}	2.041	0

and {E} are first merged at a distance 0.857. Next {A} and {C} are merged at a distance of 0.871. Using the furthest distance between observations in {D,E} and {A,C}, the two clusters are merged at a distance of 1.546 (marked at 3 on Fig. 3.12). Finally, clusters {A,C,D,E} are merged with {B} at distance 2.041 (marked 4 on Fig. 3.12).

3.3.4   Average Linkage

Unlike single or complete linkage methods, the average linkage method uses all distances between observations in two clusters to calculate a distance between the

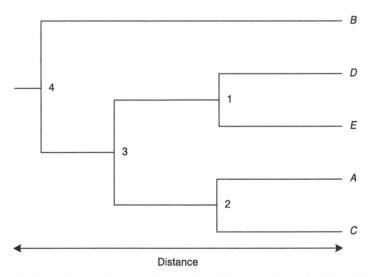

Figure 3.12   Clustering dendrogram using complete linkage agglomerative hierarchical clustering

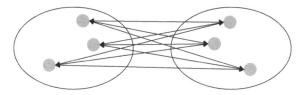

Figure 3.13   Using the average distance to decide whether to join the two groups together

clusters. The average distance between all pairs of observations, where each pair contains one observation from each cluster, is used to measure the distance between two clusters. For example, in Fig. 3.13, the nine distances between each pair of observations are summed and divided by the number of distances considered, in this case nine. The average linkage method is widely used.

Using the distance matrix in Table 3.16, it is again initially determined that the shortest distance of 0.857 is between $\{D\}$ and $\{E\}$. These two groups are discarded and replaced by the new group $\{D,E\}$. A new distance matrix is computed as shown in Table 3.23. To calculate the new distance, the average distance between each pair of observations in the two clusters is computed. For example, in computing the distance between $\{A\}$ and $\{D,E\}$, the distance of 1.036 between $\{A\}$ and $\{D\}$, and the distance of 1.547 between $\{A\}$ and $\{E\}$ are averaged. The distance between $\{A\}$ and $\{D,E\}$ is computed to be 1.291.

By examining the distance matrix in Table 3.23, it is determined that the next smallest distance is between $\{A\}$ and $\{C\}$. These individual groups are replaced by the new group $\{A,C\}$ and a new distance matrix is computed using the average linkage joining rule, as shown in Table 3.24.

Table 3.24 is then examined for the smallest distance between the remaining groups. This results in combining groups $\{D,E\}$ and $\{A,C\}$, generating a final distance matrix as shown in Table 3.25. In the last clustering step, the two remaining groups are combined at a distance of 1.504.

Figure 3.14 is a dendrogram generated by clustering the five normalized observations from Table 3.15 using agglomerative hierarchical clustering with the average linkage joining method. $\{D\}$ and $\{E\}$ are initially joined at a distance of 0.857, identified by the position of the vertical line marked as 1. Next, $\{A\}$ and $\{C\}$ are combined at a distance of 0.871, represented by the vertical line at 2. Clusters $\{D,E\}$ and $\{A,C\}$

TABLE 3.23   Distance Matrix for the Four Groups Using the Average Linkage Criteria

	$\{A\}$	$\{B\}$	$\{C\}$	$\{D,E\}$
$\{A\}$	0	2.041	0.871	1.291
$\{B\}$	2.041	0	1.185	1.395
$\{C\}$	0.871	1.185	0	0.976
$\{D,E\}$	1.291	1.395	0.976	0

TABLE 3.24 Distance Matrix for the Three Groups Using the Average Linkage Criteria

	{B}	{D,E}	{A,C}
{B}	0	1.394	1.612
{D,E}	1.394	0	1.133
{A,C}	1.612	1.133	0

TABLE 3.25 Distance Matrix for the Two Groups Using the Average Linkage Criteria

	{B}	{A,C,D,E}
{B}	0	1.504
{A,C,D,E}	1.504	0

are combined at a distance of 1.133 (shown at 3) to form cluster {A,C,D,E}. Finally {B} is combined with {A,C,D,E} at a distance of 1.504, shown at 4.

The three dendrograms generated using the single, complete, and average linkage joining rules are shown back-to-back in Fig. 3.15. In this example, they all clustered the observations in a similar manner, but at different distances. In many situations, choosing different joining rules will result in a different clustering of the observations.

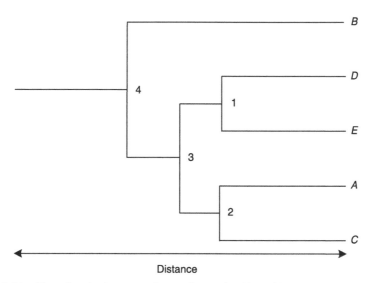

Figure 3.14 Clustering dendrogram using agglomerative hierarchical clustering with the average linkage joining rule

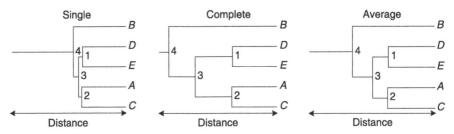

Figure 3.15   Three cluster dendrograms based on the single, complete, and average joining rules

3.3.5   Other Methods

The following summarizes two alternative hierarchical agglomerative clustering approaches:

- *Centroid*: Under this method, for each cluster, a cluster centroid is calculated as the average value across all the variables. The distance between clusters is calculated as the distance between these centroids. In Fig. 3.16, a triangle is used to illustrate the centroid of the two clusters, and the line between the two centroids illustrates the distance between the two clusters.

- *Wards*: This method for agglomerative hierarchical clustering, like the others described so far, starts with each observation as a single cluster and progressively joins pairs of clusters until only one cluster remains containing all observations. This method can operate either directly on the underlying data or on a distance matrix. When operating directly on the data, the approach is limited to data with continuous values. At each step, Wards's method uses a function to assess which two clusters to join. This function attempts to identify the cluster which results in the least variation, and hence identifies the most homogeneous group. The error sum of squares formula is used at each stage in the clustering process and is applied to all possible pairs of clusters.

3.3.6   Selecting Groups

A dendrogram describing the relationships between all observations in a data set is useful for understanding the hierarchical relationships in the data. In many situations, a discrete number of specific clusters is needed. To convert the

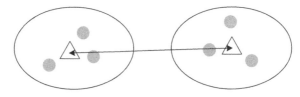

Figure 3.16   Using the centroid distance to decide whether to join the two groups together

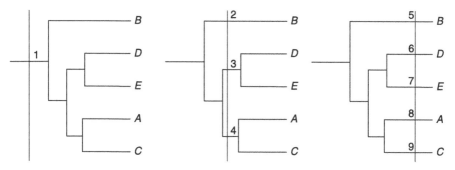

Figure 3.17   Generating one, three, and five clusters at three different distance cut-offs

dendrogram to a series of groups, a distance at which the groups are to be formed should be specified. Once this distance is selected, a line cutting through the dendrogram at the determined distance can be drawn, thus dissecting the dendrogram and forming specific clusters.

In Fig. 3.17, the same dendrogram is shown, but with three different distance cut-offs. In the first dendrogram to the left, a large distance was selected. This results in the cut-off line dissecting the dendrogram at point 1. All children to the right of point 1 would be considered to belong to the same group, and hence only one group is generated in this scenario. In the second dengrogram in the middle, a lower distance cut-off value has been set, shown by the vertical line. This line dissects the dendrogram at points 2, 3, and 4, resulting in three clusters. Where the line intersects at point 2, child B is assigned to one cluster. Another cluster is formed where the line intersects at point 3, and all the children of point 3 are assigned to a second cluster; here that is observations D and E. Similarly, there is a cluster generated through the intersection at point 4 containing observations A and C. The third dendrogram on the right has the lowest cut-off distance value, resulting in all observations being assigned to a group containing one observation. Observation B is assigned to one cluster resulting from the intersection at point 5, observation D is assigned to another cluster through the intersection at point 6, and so on.

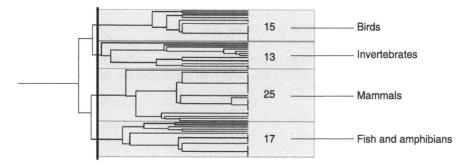

Figure 3.18   Clustering dendrogram for the animal data set, with four groups selected

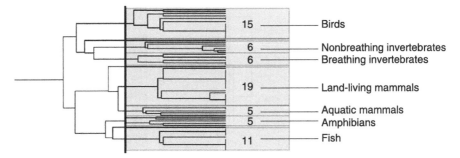

Figure 3.19   Clustering dendrogram for the animal data set

Figures 3.18 and 3.19 show the clustering dendrogram generated for the data set of 70 animals described in Section 3.1, using an agglomerative hierarchical cluster, the Gower distance, and the average linkage joining rule. Figure 3.18 shows the clustering dendogram for this data set when a cut-off has been set such that four groups are selected. These four groups are the same groups as shown in Fig. 3.1. Moving the cut-off to generate 10 groups is presented in Fig. 3.19, and these groups correspond to the animal clusters in Fig. 3.2.

The distance at which the clusters are to be generated will be based on the specific problem being addressed. Generally, setting higher distance cut-offs will result in smaller numbers of clusters, with greater diversity within each individual cluster. Setting lower distance cut-offs results in greater numbers of more homogeneous groups.

3.4   PARTITIONED-BASED CLUSTERING

3.4.1   Overview

Partitioned-based methods generate a specified number of clusters, usually referred to as k. This number is set prior to performing any clustering where k must be at least 2 and less than the total number of observations in the data set. Partitioned-based clustering is computationally more efficient for grouping observations than agglomerative methods and can be applied to large data sets. However, this method is somewhat sensitive to noise and outliers in the data, and the number of clusters must be specified at the start. Some familiarity with the data set and the problem is helpful in order to identify a value for k. Alternatively, running the clustering multiple times with different values for k, and then selecting the value which generates the optimal clustering, is also used.

3.4.2 k-Means

k-Means is one partitioned-based approach to clustering a data set. Because one of the clustering steps is to calculate means for variables across observations within clusters, this method is limited to use with continuous variables. Having specified a number of clusters (k), initially k observations, referred to as *seeds*, are selected. This initial assignment may be randomly generated, or it may be selected with some knowledge of the content. Following the selection of the k cluster seeds, all other observations in the data set are assigned to the seed closest to it, based on the distance between the observation and each of the seeds. Any distance measure, such as those discussed in Section 3.2.2, can be used to measure the distance between pairs of observations. This generates an initial grouping of the data into k clusters.

The next step is to optimize the groupings. A mean value, or *centroid*, is calculated for each of the groups across all the variables. All observations are then reassigned to the cluster whose centroid they are closest to. At this point an assessment should be made as to whether the grouping is optimal or not. If the assessment is that the grouping is optimal, then the clustering is complete. If not, the process of calculating the cluster means and then reassigning all observations is repeated. The assessment of an optimal solution is based on whether any observations are reassigned to a different cluster or based on an error criterion (*Err*):

$$Err = \sum_{i=1}^{k} \sum_{x \in C_i} d_{x,\mu(C_i)}$$

This criterion is based on looking at observations (x) in all k clusters, where $\mu(C_i)$ is the mean for the cluster. When the error criterion becomes less than a predetermined value, the clustering stops.

In the following example, the normalized data shown in Table 3.26 is clustered into three groups using k-means clustering. The euclidean distance is used to calculate the distance between observations. Initially, one observation is assigned to each of the three clusters. In this example, the assignment is done randomly. Observation B is assigned to cluster 1, observation C to cluster 2, and observation E to cluster 3. Next, all remaining observations are placed in the cluster closest to them. Observation B is placed into cluster 1; observations A and C are placed into cluster 2; and observations D and E are placed into cluster 3. Once the initial clusters have been populated, the mean for each cluster (centroid) is calculated, as shown in Table 3.27. Each observation is then re-examined to determine whether it belongs to a different group. In this example, this examination results in no changes to the group assignment, and hence these clusters are the final grouping.

The k-means clustering, described in this section, is dependent to some extent on the initial assignment of observations, especially if the initial assignment of

TABLE 3.26   Table of Normalized Observations

	Percentage body fat	Weight	Height	Chest	Abdomen
A	0.0	0.0	0.0	0.0	0.0
B	1.0	1.0	0.409	1.0	1.0
C	0.513	0.375	0.136	0.441	0.377
D	0.00932	0.347	0.955	0.178	0.050
E	0.585	0.681	1.0	0.520	0.562

TABLE 3.27 Calculated Mean Values (or Centroids) and Assignment of Observations to the Three Clusters

	Percentage body fat	Weight	Height	Chest	Abdomen
Cluster 1					
B	1.0	1.0	0.409	1.0	1.0
Mean	1.0	1.0	0.409	1.0	1.0
Cluster 2					
A	0.0	0.0	0.0	0.0	0.0
C	0.513	0.375	0.136	0.441	0.377
Mean	0.256	0.187	0.0682	0.220	0.188
Cluster 3					
D	0.00932	0.347	0.955	0.178	0.050
E	0.585	0.681	1.0	0.520	0.562
Mean	0.339	0.514	0.977	0.349	0.306

observations is made at random. It is therefore worthwhile repeating the clustering and comparing the different runs, selecting the most consistent grouping.

3.4.3 Worked Example

To illustrate k-means clustering, a data set of wines will be used that are characterized by a number of chemical properties (http://archive.ics.uci.edu/ml/datasets/Wine), with a portion of the data shown in Table 3.28.

    To quickly understand the content of the table, the data was clustered using k-means clustering. The number of clusters (k) was set to three, and the euclidean distance was used as a distance measure between observations. Figure 3.20 summarizes the resulting clustering. For each of the three clusters, the mean values across all variables (centroids) for the cluster are displayed in a row of the table, along with a count

TABLE 3.28 Data Set of Wines

Malic acid	Ash	Alkalinity of ash	Magnesium	Total phenols	Nonflavanoids
14.23	1.71	2.43	15.6	127	3.06
13.2	1.78	2.14	11.2	100	2.76
13.16	2.36	2.67	18.6	101	3.24
14.37	1.95	2.5	16.8	113	3.49
13.24	2.59	2.87	21	118	2.69
14.2	1.76	2.45	15.2	112	3.39
14.39	1.87	2.45	14.6	96	2.52
14.06	2.15	2.61	17.6	121	2.51
14.83	1.64	2.17	14	97	2.98

Centroid values

Cluster ID	Malic acid	Ash	Alkalinity of ash	Magnesium	Total phenols	Nonflavanoids	Count
Cluster 1	13.1	3.62	2.44	21.5	97.7	1.06	54
Cluster 2	12.2	1.7	2.24	20	94.8	1.98	64
Cluster 3	13.7	1.86	2.44	17.1	107	2.95	60

Figure 3.20   Summary of three clusters generated from k-means clustering

of the number of observations. Figures 3.21, 3.22, and 3.23 provide a histogram matrix for each of the clusters, with the observations in the selected cluster displayed as dark shaded areas on the histograms.

3.4.4   Miscellaneous Partitioned-Based Clustering

Two alternative partitioned-based clustering methods are:

- *k-modes*: This clustering approach can be used with categorical data, and it operates similarly to k-means. Where k-means uses a mean function to compute the cluster centroid, k-modes uses a mode function for this computation.

- *k-medoids*: This clustering approach does not use a mean centroid, but it instead uses a representative observation as the cluster centroid. This approach helps overcome issues with outliers that can skew results generated with the k-means method.

Centroid values

Cluster ID	Malic acid	Ash	Alkalinity of ash	Magnesium	Total phenols	Nonflavanoids	Count
Cluster 1	13.1	3.62	2.44	21.5	97.7	1.06	54
Cluster 2	12.2	1.7	2.24	20	94.8	1.98	64
Cluster 3	13.7	1.86	2.44	17.1	107	2.95	60

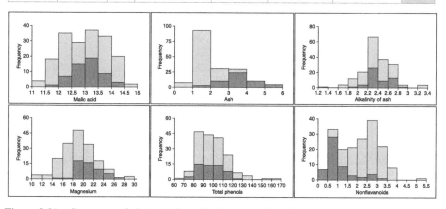

Figure 3.21   Summary of cluster 1 from the k-means clustering

Centroid values

Cluster ID	Malic acid	Ash	Alkalinity of ash	Magnesium	Total phenols	Nonflavanoids	Count
Cluster 1	13.1	3.62	2.44	21.5	97.7	1.06	54
Cluster 2	12.2	1.7	2.24	20	94.8	1.98	64
Cluster 3	13.7	1.86	2.44	17.1	107	2.95	60

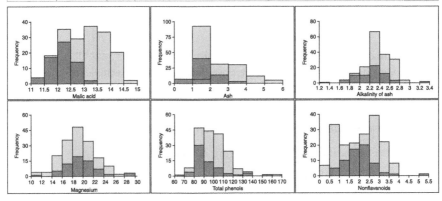

Figure 3.22   Summary of cluster 2 from the *k*-means clustering

Centroid values

Cluster ID	Malic acid	Ash	Alkalinity of ash	Magnesium	Total phenols	Nonflavanoids	Count
Cluster 1	13.1	3.62	2.44	21.5	97.7	1.06	54
Cluster 2	12.2	1.7	2.24	20	94.8	1.98	64
Cluster 3	13.7	1.86	2.44	17.1	107	2.95	60

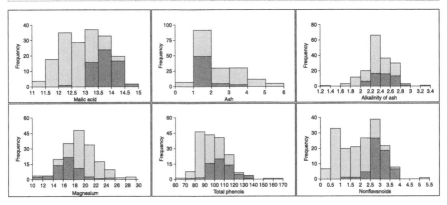

Figure 3.23   Summary of cluster 3 from the *k*-means clustering

3.5 FUZZY CLUSTERING

3.5.1 Overview

With partitioned-based clustering (also referred to as "hard" or "crisp" clustering), each observation belongs to a single cluster, there are no empty clusters, and the number of clusters needs to be specified prior to clustering. Like partitioned-based approaches, fuzzy clustering requires the number of clusters to be specified prior to clustering. However, unlike partitioned-based clustering, each observation belongs to all clusters. Accompanying each observation is a number that reflects the degree to which the observation belongs to each cluster. This membership is calculated using a *membership function* whereby the degree of association is reflected in a number between 0 and 1. Lower numbers indicate that the observation is marginally associated with the cluster, and higher numbers show the observation is strongly associated with a particular cluster. The results of fuzzy clustering are a membership matrix composed of k columns corresponding to the number of clusters, and n observations, as shown in Table 3.29. This membership matrix was calculated from the observations in Table 3.26.

In the example in Table 3.29, five observations (A, B, C, D, and E) have been clustered into three groups (cluster 1, cluster 2, and cluster 3). The degree of association with each cluster is shown. Observation A has almost no association with cluster 1 and cluster 2; however, it has a strong association with cluster 3. Observation C is most strongly associated with cluster 3, but it also has some level of association with cluster 1 and cluster 2. For each observation, the sum of all cluster associations for each observation is 1.

3.5.2 Fuzzy *k*-Means

Like k-means clustering, *fuzzy k-means* clustering requires that the number of clusters must be specified prior to clustering. This number, often referred to as k, must be greater than one and less than the total number of observations in the data set. Another parameter is often referred to as the degree of fuzziness (q), and it also must be specified prior to clustering. There are two important functions that are required for fuzzy k-means clustering: the membership function and the function to calculate the cluster centroid. These calculations are repeatedly used in the clustering process to optimize the results.

TABLE 3.29 Example Membership Matrix

Label	Cluster 1	Cluster 2	Cluster 3
A	0.0272	0.00911	0.964
B	0.00253	0.996	0.00147
C	0.336	0.179	0.485
D	0.803	0.045	0.152
E	0.705	0.199	0.0962

The membership function defines the degree of association between an observation and a specific cluster. The function is used to calculate an entry for each observation in each cluster in order to construct the entire membership matrix. The following formula is used:

$$m_{ij} = \frac{d_{ij}^{-2/(q-1)}}{\sum_{l=1}^{k} d_{il}^{-2/(q-1)}}$$

In this formula, a membership score (m_{ij}) is calculated that reflects the association of observation i to cluster j. The value of i can take any value between one and the total number of observations in the data set (n), and j can take any value between one and the number of clusters (k). d_{ij} is the distance between observation i and cluster centroid j, and d_{il} is the distance between the observation i and each of the cluster centroids l.

The optimal value for the fuzziness parameter q is dependent on the specific data set being analyzed, with typical values ranging between 1.5 and 9. Higher values for q result in increasingly fuzzier clustering.

The formula for computing the cluster centroid is:

$$C_j = \frac{\sum_{i=1}^{n} m_{ij}^q x_i}{\sum_{i=1}^{n} m_{ij}^q}$$

In this formula, a value for the centroid C_j for cluster j is calculated. The formula makes use of the membership function value for observation i, and its association with cluster j, raised to the power of the fuzziness parameter $q(m_{ij}^q)$. The observation values are incorporated (x_i).

Initially, a number of centroid values are randomly assigned using random observations or random values. Using these centroid values, an initial membership matrix is calculated, followed by a calculation of the centroids based on this matrix. The process of recalculating the membership matrix, followed by a recalculation of the centroid, is repeated until the matrix is determined to be optimized. The membership matrix is considered optimized when the largest difference of any two corresponding cells of the previous membership matrix and the newly computed membership matrix is less than some error threshold. At that point, the clustering process ends and the optimized matrix is used as the clustering results.

3.5.3   Worked Examples

The normalized table of values shown in Table 3.30 will be used to illustrate fuzzy k-means clustering. In this example, the euclidean distance is used to calculate distances between observations, the fuzziness value is given a value of 2, and three clusters will be generated. Prior to clustering, the predefined termination value is set to 0.005.

Initially, a centroid is randomly assigned to each of the three clusters (see Table 3.31). Then a membership matrix is calculated based on the membership

TABLE 3.30   Example Table of Normalized Data

	Percentage body fat	Weight	Height	Chest	Abdomen
A	0.0	0.0	0.0	0.0	0.0
B	1.0	1.0	0.409	1.0	1.0
C	0.513	0.375	0.136	0.441	0.377
D	0.00932	0.347	0.955	0.178	0.050
E	0.585	0.681	1.0	0.520	0.562

TABLE 3.31   Initial Random Centroids

	Percentage body fat	Weight	Height	Chest	Abdomen
Centroid 1	0.543	0.187	0.573	0.812	0.294
Centroid 2	0.922	0.482	0.903	0.551	0.197
Centroid 3	0.925	0.281	0.098	0.220	0.282

function. In this example, the membership matrix makes use of this assigned centroid and uses the euclidean distances to the observations (see Table 3.32). The centroid function is now recalculated, using this membership matrix (see Tables 3.33 and 3.34). All elements of this membership matrix are compared to the previously calculated membership matrix, and the element with the maximum difference is determined. This value is referred to as the maximum error, which here is 0.234. Because the error is greater than the predetermined cutoff value, which is 0.005 in this example, the process continues.

TABLE 3.32   Initial Membership Matrix

	Cluster 1	Cluster 2	Cluster 3
A	0.341	0.213	0.446
B	0.349	0.365	0.286
C	0.331	0.152	0.517
D	0.405	0.368	0.228
E	0.291	0.574	0.135

TABLE 3.33   First Updated Centroid

	Percentage body fat	Weight	Height	Chest	Abdomen
Centroid 1	0.384	0.465	0.514	0.408	0.367
Centroid 2	0.509	0.620	0.774	0.509	0.502
Centroid 3	0.372	0.344	0.223	0.353	0.316

TABLE 3.34   First Updated Membership Matrix

	Cluster 1	Cluster 2	Cluster 3
A	0.341	0.213	0.446
B	0.349	0.365	0.286
C	0.331	0.152	0.517
D	0.405	0.367	0.228
E	0.291	0.574	0.135

The centroid is recalculated and a new membership function is determined, followed by an evaluation of the maximum error. The maximum error is now 0.128, which is considerably less than the initial error, yet still greater than the 0.005 cutoff value. The process of recalculating the centroid and the membership matrix is continued until the maximum error value is less than 0.005. On the thirty-fifth iteration, the maximum error is recorded as 0.00465. Since the error is now less than 0.005, the process finishes and the final centroid values and the membership matrix represent the final clustering results, as shown in Tables 3.35 and 3.36, respectively.

The final membership matrix shows how each observation belongs to each of the three clusters. Observation A is primarily associated with cluster 3, and observation B is primarily associated with cluster 2. Observations C, D, and E are not as clearly associated with a single cluster. Observation C is primarily associated with cluster 3; however, it is also associated with cluster 1. Observations D and E are primarily associated with cluster 1, but they have levels of association with the remaining clusters.

TABLE 3.35   Thirty-fifth Updated Centroids

	Percentage body fat	Weight	Height	Chest	Abdomen
Centroid 1	0.283	0.482	0.898	0.337	0.282
Centroid 2	0.968	0.968	0.424	0.964	0.963
Centroid 3	0.106	0.0857	0.0530	0.0941	0.0794

TABLE 3.36   Thirty-fifth Updated Membership Matrix

	Cluster 1	Cluster 2	Cluster 3
A	0.0272	0.00911	0.964
B	0.00253	0.996	0.00147
C	0.336	0.179	0.485
D	0.802	0.0452	0.152
E	0.705	0.199	0.0962

TABLE 3.37 **Example Flowers from the Iris Data Set**

ID	Sepal length	Sepal width	Petal length	Petal width
A	6.3	3.3	4.7	1.6
B	7.4	2.8	6.1	1.9
C	4.7	3.2	1.3	0.2
D	5.1	3.5	1.4	0.2
E	4.9	3	1.4	0.2
F	4.6	3.1	1.5	0.2
G	5	3.6	1.4	0.2
H	5.4	3.9	1.7	0.4
...	...	...	...	...

To further illustrate fuzzy clustering, the Iris data set of 150 flowers (http://archive.ics.uci.edu/ml/datasets/Iris) is used, where sepal length, sepal width, petal length, and petal width have been measured. Table 3.37 details a few examples from the data set.

Fuzzy clustering is performed on the data where the number of clusters is set to three, and the euclidean distance is used to assess the distance between observations. Table 3.38 shows the results of the clustering, where each observation is assigned a number between zero and one for each of the three clusters: cluster 1, cluster 2, and cluster 3.

Figure 3.24 contains three scatterplots, where *sepal length* and *petal length* have been plotted on identical axes. On each of the scatterplots, the 50 highest-ranked observations for each of the clusters are highlighted. The ranking is based on the cluster membership function shown in Table 3.38. Generally, cluster 1 is centered in the top right of the scatterplot, cluster 2 is centered in the middle, and cluster 3 is centered near the bottom left. Observations *A*, *B*, and *C* from Table 3.38 are shown on the scatterplot. Observation *A* is located between cluster 1 and cluster 2 on Fig. 3.24, and this is reflected by the membership function values. Observations *B* and *C* are most strongly associated with cluster 1 and cluster 3, respectively, as shown in Table 3.38 and Fig. 3.24.

TABLE 3.38 **Fuzzy Clustering Results from the Iris Data Set**

ID	Sepal length	Sepal width	Petal length	Petal width	Cluster 1	Cluster 2	Cluster 3
A	6.3	3.3	4.7	1.6	0.47	0.48	0.05
B	7.4	2.8	6.1	1.9	0.82	0.15	0.03
C	4.7	3.2	1.3	0.2	0.01	0.02	0.97
D	5.1	3.5	1.4	0.2	0.00	0.00	0.99
E	4.9	3	1.4	0.2	0.02	0.05	0.93
F	4.6	3.1	1.5	0.2	0.02	0.04	0.94
G	5	3.6	1.4	0.2	0.01	0.01	0.98
H	5.4	3.9	1.7	0.4	0.05	0.09	0.87
...	...	...	...	...	...	...	...

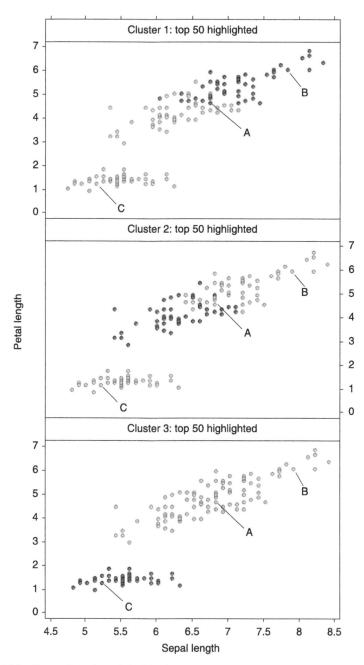

Figure 3.24 Fuzzy clustering of the Iris data set

3.6 SUMMARY

The starting point for any clustering method is a *distance matrix* or *dissimilarity matrix* (*D*), which is a square *n*-by-*n* matrix, where *n* is the number of observations in the data set to be clustered. This distance matrix has the following format:

$$D = \begin{bmatrix} 0 & d_{1,2} & d_{1,3} & \cdots & d_{1,n} \\ d_{2,1} & 0 & d_{2,3} & \cdots & d_{2,n} \\ d_{3,1} & d_{3,2} & 0 & \cdots & d_{3,n} \\ \cdots & \cdots & \cdots & \cdots & \cdots \\ d_{n,1} & d_{n,2} & d_{n,3} & \cdots & 0 \end{bmatrix}$$

Tables 3.39 and 3.40 summarize the distance measures covered in this chapter. Table 3.41 summarizes the clustering approaches described in this chapter.

TABLE 3.39 Distance Measures Covered in This Chapter

Numeric	Binary
Euclidean: $d_{p,q} = \sum_{i=1}^{n} \sqrt{(p_i - q_i)^2}$	Simple matching: $d_{p,q} = 1 - \dfrac{a+d}{a+b+c+d}$
Square euclidean: $d_{p,q} = \sum_{i=1}^{n} (p_i - q_i)^2$	Jaccard: $d_{p,q} = \dfrac{b+c}{a+b+c}$
Manhattan: $d_{p,q} = \sum_{i=1}^{n} |p_i - q_i|$	Russell and Rao Y: $d_{p,q} = 1 - \dfrac{a}{a+b+c+d}$
Maximum: $d_{p,q} = \max |p_i - q_i|$	Dice: $d_{p,q} = \dfrac{b+c}{2a+b+c}$
Minkowski: $d_{p,q} = \sqrt[\lambda]{\sum_{i=1}^{n} |p_i - q_i|^{\lambda}}$	Rogers and Tanimoto: $d_{p,q} = \dfrac{2(b+c)}{a+2(b+c)+d}$
Canberra: $d_{p,q} = \sum_{i=1}^{n} \dfrac{|p_i - q_i|}{(|p_i| + |q_i|)}$	
Mahalanobis: $d_{p,q} = \sqrt{(p-q)S^{-1}(p-q)^T}$	

TABLE 3.40 Additional Distance Measures

Mixed	Variables
Gower: $d_{p,q} = \sqrt{\dfrac{\sum_{i=1}^{n} w_i d_i^2}{\sum_{i=1}^{n} w_i}}$	Correlation coefficient: $r = \dfrac{\sum_{i=1}^{n} (x_i - \bar{x})(y_i - \bar{y})}{(n-1)s_x s_y}$

TABLE 3.41 Summary of Clustering Approaches

	Advantages	Disadvantages	Features
Hierarchical	Shows hierarchical relationships Does not require selection of cluster numbers upfront	Slow to compute Cannot be applied to large data sets	All observations are in a single cluster
Partitioned	Fast to compute Works with large data sets	Need to specify the number of clusters upfront Does not show relationship among clusters	Each observation belongs to a single cluster
Fuzzy	Fast to compute Works with large data sets	Need to specify the number of clusters upfront	All observations are in all clusters, but to various degrees

3.7 FURTHER READING

The following sources provide additional general information on clustering: Gan and Wu (2007), Mirkin (2005), Everitt et al. (2001), Kaufmann and Rousseeuw (2005), Han and Kamber (2005), and Tan et al. (2005). A number of methods have been developed to aid in selecting the number of clusters, which is especially useful with clustering approaches such as k-means clustering, where a value for k must be set prior to clustering: Milligan and Cooper (1985), Kelley et al. (1996), Kyrgyzov et al. (2007), Salvador and Chan (2004), and Aldenderfer and Blashfield (1984).

Other clustering approaches include divisive hierarchical clustering, self-organizing maps (SOMs), grid-based clustering, model-based clustering, and density-based clustering. Divisive hierarchical clustering operates in the reverse direction to agglomerative hierarchical clustering, starting with all observations in a single cluster, and continually dividing the set until each observation is in a single cluster. For more information on divisive hierarchical clustering, see Gan and Wu (2007) and Tan et al. (2005). A SOM or Kohenen network is a type of neural network that is often used to take high dimensional data and map it onto fewer dimensions so that it can be displayed. This method is described in Kohonen (1990, 2001), Hastie et al. (2003), Gan and Wu (2007), and Han and Kamber (2005). Grid-based clustering does not utilize the individual observations for clustering, but uses a finite number of cells resulting in faster execution, see Gan and Wu (2007), and Han and Kamber (2005). Model-based clustering attempts to generate models from the data, with each model representing a cluster (Gan and Wu, 2007; Han and Kamber, 2005). Density-based clustering is used to identify arbitrary shaped dense clusters that are separated by less dense regions. The approach handles noise and outliers well, and there is no requirement to specify the number of clusters prior to clustering. Gan and Wu (2007) and Tan et al. (2005) provide an overview of this approach. A number of references provide a greater level of detail concerning methods for assessing individual clusters such as Tan et al. (2005) and Gan and Wu (2007).

CHAPTER 4

PREDICTIVE ANALYTICS

4.1 OVERVIEW

4.1.1 Predictive Modeling

Predictive analytics refers to a series of techniques concerned with making more informed decisions based on an analysis of historical data. These methods are used throughout the industries of science and business. For example, pharmaceutical companies use these techniques to assess the safety of potential drugs before testing them in human clinical trials, and marketing organizations use these models to predict which customers will buy a specific product.

This section will use the following example to illustrate the process of building and using predictive models. A telecommunications company wishes to identify and prioritize all customers they believe are likely to change to another service provider. In an attempt to avoid losing business, the company will offer new products or new service options to these customers. Over the years, the organization collected customer data monthly that included decisions to switch to another service provider. Table 4.1 is an example of the information collected on customers, where each observation relates to a specific customer in one month. The data includes a variable, *churn*, where a 1 indicates a switch to a new service provider and a 0 indicates a continuation of service. The table also includes the specific month the observation was recorded (*month*), the age of the customer (*age*), the annual income of the customer (*income*), the number of months they have been a customer (*customer length*), the gender of the customer (*gender*), the number of calls made that month (*monthly calls*), and the number of calls made to the customer service department (*service requests*). This data will be used to build a model attempting to understand any relationships between (1) the customer data collected and (2) whether the customer changes service provider.

The telecommunications company would like to use the model in their marketing department to identify customers likely to switch. Since the company has limited resources, it would also like to prioritize the customers most likely to switch. The company has data for the current month, which is shown in Table 4.2. The data for the current month is the same type of data used to build the model, except that *churn* is currently unknown. The model created from the historical data can then be used with the new data to make a series of predictions. Table 4.3 is an example of a table where two columns have been added for the values generated by the

TABLE 4.1 Example of Telecommunications Data Used to Build a Model

ID	Month	Age	Income	Customer length	Gender	Monthly calls	Service requests	Churn
A	January	45	$72k	36	Female	46	1	0
B	March	27	$44k	24	Male	3	5	1
C	July	51	$37k	47	Male	52	0	0
D	February	17	0	16	Female	62	1	1
E	December	45	$63k	63	Female	52	0	0
F	October	24	$36k	24	Male	72	1	0
G	March	39	$48k	5	Male	36	0	0
H	June	46	$62k	17	Male	1	0	1
...	...	...	...	...	...	...	...	...

TABLE 4.2 Data Collected on Customers for the Current Month

ID	Month	Age	Income	Customer length	Gender	Monthly calls	Service requests	Churn
a	May	52	$84k	52	Female	52	0	?
b	May	26	$28k	14	Male	12	2	?
c	May	64	$59k	4	Male	31	1	?
...	...	...	...	...	...	...	...	...

model. *Predicted churn* indicates whether that particular customer is likely to switch services (*predicted churn* = 1) or not (*predicted churn* = 0) this month. *Churn probability* reflects the likelihood, or probability, that a particular customer will be assigned a *predicted churn* value of 1. This table can then be sorted by the probability column, allowing the marketing team to focus on those customers most likely to switch.

Figure 4.1 illustrates the general process of generating and using a prediction model. Data is used to generate a model that predicts a specific response variable from a set of independent variables. In the telecommunications example, the independent variables are *age, income, gender, customer length, monthly calls,* and *service requests*. These independent variables are used in concert with the response variable,

TABLE 4.3 Customers Predicted to Change Services this Month, and a Measure of the Likelihood of Switching

ID	Month	Age	Income	Customer length	Gender	Monthly calls	Service requests	Predicted churn	Churn probability
a	May	52	$84k	52	Female	52	0	0	0.33
b	May	26	$28k	14	Male	12	2	1	0.74
c	May	64	$59k	4	Male	31	1	1	0.88
...	...	...	...	...	...	...	...	...	...

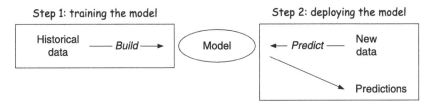

Figure 4.1   Process of generating and using a prediction model

or *churn* in this example. The model captures relationships between the independent variables and the response variable. Models are built in an initial training step where the model learns from the historical data. Once a model has been built, a new data set containing the same independent variables can be used with the model to calculate predictions from the captured relationships. In Fig. 4.1, this is identified as the second step, where the model is deployed and being used to predict a value for a response variable.

Building predictive models assumes the same exploratory data analysis as discussed in the previous chapters has been done. (The first three chapters discussed the process and methods such as data visualization, descriptive statistics, and clustering that help understand the content.)

For the model to capture important relationships, the data set must have a balanced mixture of positive and negative examples. In the telecommunications example, this means including observations where the customer did and did not switch. It will be hard to build an objective model without a good proportion of both cases because the model may falsely identify certain characteristics that are merely common to all customers, not specifically to those that switch services.

Preparing and selecting which variables should be used with a model is a critical phase in the process. Prioritizing the variables to be considered is important, especially when the data set has many variables. One way to prioritize is to look at the relationships between the potential independent variables and the target response variable to see if there is any relationship. Using methods such as a hypothesis test, a chi-square test, or a matrix of correlation coefficients can help to prioritize the variables. There should be no strong relationships between the independent variables included in the model because related variables are redundant, or worse, they may introduce problems when building models. Knowledge of the specific subject matter and an understanding of how the model is going to be deployed is also often critical to making choices about what variables to include. For example, if collecting or measuring a particular variable's values is too costly, then this variable should be excluded despite its potential utility.

The next major factor that affects the model-building approach is the type of response variable involved, that is, whether the variable is categorical or continuous. If the response variable is categorical, then a *classification* modeling approach should be used. These include logistic regression, discriminant analysis, naive Bayes, *k*-nearest neighbors (*k*NN), classification trees, and neural networks. If the response variable is continuous, then a *regression* modeling approach should be considered. These include linear regression, *k*-nearest neighbors, regression trees,

TABLE 4.4 Summary of Different Modeling Approaches

Method	Model type	Independent variables	Comments
Linear regression	Regression	Any numeric	Assumes a linear relationship Easy to explain Quick to build
Discriminant analysis	Classification	Any numeric	Assumes the existence of mutually exclusive groups with common variances
Logistic regression	Classification	Any numeric	Will calculate a probability Easy to explain
Naive Bayes	Classification	Only categorical	Requires a lot of data
Neural networks	Regression or classification	Any numeric	Black box model
kNN	Regression or classification	Any numeric	Difficult to explain results Handles noise well Handles nonlinear relationships
CART	Regression or classification	Any	Explanation of reasoning through use of a decision tree

and neural networks. A few approaches may be used with categorical and continuous responses. All approaches have advantages and disadvantages, and the different approaches often require the data to conform to certain assumptions. The major approaches are summarized in Table 4.4.

In selecting which approach to use, other practical considerations may need to be considered. In the telecommunication example, the marketing department needed a classification model to predict the binary response variable *churn*; however, the department also wanted to prioritize the results so they can focus their efforts on customers most likely to switch. In this situation, the logistic regression method might be a good candidate since it generates a prediction for the binary variable *churn* and calculates the probability of switching.

Different modeling approaches operate in different ways and selecting the best method requires an understanding of the data and how the different methods operate. Figure 4.2 illustrates different types of approaches. In chart A, a single independent variable is plotted against a single response variable. Because a linear relationship exists between the two variables, they can be modeled using a linear regression method. In contrast, chart B indicates a nonlinear relationship where the data either needs to be transformed or needs to be used with a method that is capable of modeling these relationships. In chart C, the data used to build the model was divided into regions based on specific values or ranges of the independent variables. In this example, two independent variables are used to illustrate how the data can be divided into regions (no response variable is shown). When making a prediction, the approach assigns an observation into a specific region based on values or ranges for the independent variables, and a prediction is made based on training data in the region, such

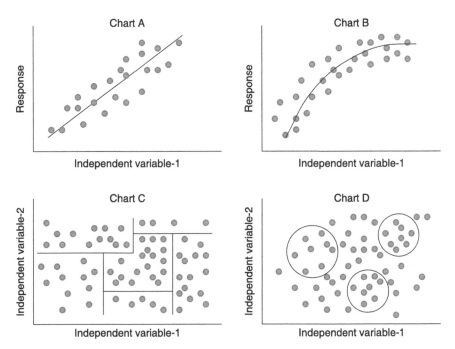

Figure 4.2   Illustrations of different regression modeling approaches

as the average value. Techniques such as regression trees make predictions in this manner. Another approach is illustrated in chart D, where similar observations from the training data to the observation to be predicted are identified and a prediction is made based on the average response values from this set. kNN uses this approach to make predictions.

Figure 4.3 illustrates a number of classification approaches. In chart A, the independent variable space is characterized by grouping similar observations. A prediction is made by deciding what similar observations are present in the training set, and then using the mode response value from these observations as the predicted classification. Again, the kNN method is an example of this approach. Similarly, the training data could have been divided into specific regions based on a good classification

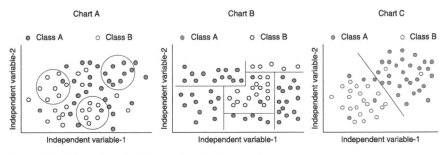

Figure 4.3   Illustration of different classification modeling approaches

of the data, and predictions may be made by determining where in this space a test observation should be assigned. Classification trees are an example of this approach. Another approach is to determine whether there are boundaries between the classes, as illustrated in chart C. These boundaries allow for classification, and methods such as discriminant analysis will model the data based on this type of approach.

Modeling the data using different approaches and fine-tuning the parameters used within each method will lead to models with greater predictive accuracy. Understanding how the individual approaches operate will help in this optimization process. Building multiple models and assessing them with a consistent methodology will help to select the best one for the particular problem being addressed. Care must be taken to build the model such that the relationships are general enough to make good predictions beyond the specific examples in the training data. In addition, the bounds of the model should be characterized based on the types and ranges of data used to build the model or based on general population characteristics or assumptions.

This chapter describes a series of multivariate approaches for building and testing predictive models. Section 4.1.2 describes methods for dividing data sets into sets for training and testing the model, thus ensuring objective testing of the model. Different metrics for assessing models are provided in Sections 4.1.3−4.1.7. These methods differ based on the type of the response variable, that is, whether the variable is continuous, categorical, or binary. Prior to building any model, it is important to understand and select variables to use as independent variables in the model. In Section 4.2, a technique referred to as *principal component analysis* is described. This technique helps in selecting variables or determining derived variables to use in any model. Sections 4.3−4.7 describe a series of widely used modeling approaches, including multiple linear regression, discriminant analysis, logistic regression, and naive Bayes.

4.1.2   Testing Model Accuracy

In order to assess which predictive data mining approach is most promising, it is important to assess the various options in a way that is objective and consistent. Evaluating the different approaches also helps set expectations about performance levels for a model ready to be deployed. In evaluating a predictive model, different data sets should be used to build the model and to test the performance of the model. Using different data ensures that the model has not overfitted the training data. The following approaches are commonly used to achieve this:

- *Test data*: Before the data set is used to train any models, a set of data selected randomly is set aside for the sole purpose of testing the quality of the results, such as one-third of the data set. These observations will not be used in building the model, but they will be used with any built model to test the model's predictive performance. It should be noted that, in the ideal case, the test set is only used for model assessment. However in practical situations, there may not be enough data available.

- *Cross-validation*: The same set of observations can be used for both training and testing a model, but not at the same time. In the cross-validation approach, a percentage of the data set is assigned for test purposes. Then, multiple training and validation set combinations are identified to ensure that all observations will be a member of one validation set, and hence there will be a prediction for each observation. The assignment to the training and validation sets is random, and all validation sets will be mutually exclusive and approximately the same size. As an example, if the validation set size percentage is 10%, one-tenth of the data set will be set aside for testing and the remaining nine-tenths used for training the model. Under this scenario, 10 models must be built to ensure that all observations will be tested.

4.1.3 Evaluating Regression Models' Predictive Accuracy

The following section discusses methods used to assess the accuracy of a regression model, that is, a model built to predict a continuous variable. One effective way to visualize the accuracy of the model is to draw a scatterplot of the actual response values against the predicted response values. In Fig. 4.4 a model was built with the actual response values (y) plotted against the predicted values ($\hat{y}$). The reference line indicates where values would lie if the model made a perfect prediction, that is, when the predicted values are equal to the actual values. A good model has points close to the line, like the model displayed in Fig. 4.4.

A number of methods can be used to assess the model. The *error* or *residual* refers to the difference between the actual response value (y_i) and the predicted response value ($\hat{y}_i$). To quantify the error over the entire test set, the squared or absolute error is used, thus avoiding using a negative number which would bias the overall error evaluation. The *mean square error* and the *mean absolute error* sum these errors and divide the sum by the number of observations (n). Both values provide a good indication of the overall error level of the model.

$$\text{mean square error} \qquad \frac{\sum_{i=1}^{n} (\hat{y}_i - y_i)^2}{n}$$

$$\text{mean absolute error} \qquad \frac{\sum_{i=1}^{n} |\hat{y}_i - y_i|}{n}$$

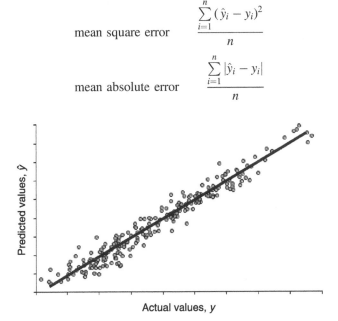

Figure 4.4   Scatterplot showing the actual values plotted against predicted values

Two additional approaches, *relative square error* and *relative absolute error*, normalize the overall error based on using the mean value ($\bar{y}$) as a simple prediction:

$$\text{relative square error} \quad \frac{\sum\limits_{i=1}^{n}(\hat{y}_i - y_i)^2}{\sum\limits_{i=1}^{n}(y_i - \bar{y})^2}$$

$$\text{relative absolute error} \quad \frac{\sum\limits_{i=1}^{n}|\hat{y}_i - y_i|}{\sum\limits_{i=1}^{n}|y_i - \bar{y}|}$$

The *correlation coefficient* is a measure of the linear relationship between two continuous variables. In this situation, the variables are the actual values and the predicted values. The resulting values are always between -1 and $+1$ where strong positive linear relationships are signified by values close to $+1$, strong negative linear relationships are close to -1, and values close to 0 indicate a lack of any linear relationship. When this value is squared, the resulting range will be between 0 and 1. The equation uses the average value of both the actual values ($\bar{y}$), and the predicted values ($\bar{\hat{y}}$), as well as the standard deviation of the actual value (s_y), and the standard deviation of the predicted values ($s_{\hat{y}}$).

$$\text{correlation coefficient} \quad \frac{\sum\limits_{i=1}^{n}(y_i - \bar{y})(\hat{y}_i - \bar{\hat{y}})}{(n-1)s_y s_{\hat{y}}}$$

Figure 4.5 displays the results from three models: models A, B, and C. The figure shows three different scatterplots of a model's predictions against the actual values. The closer the predicted values are to the actual values for the entire data set, the better the model is. Model A is the least predictive of the three, with model B providing a greater level of prediction, and model C showing the best level of accuracy.

The metrics described for assessing predictive accuracy are calculated for the three models and shown in Table 4.5. The first four values (mean square error, mean absolute error, relative square error, and relative absolute error) all have values that decrease with improved predictive accuracy. The correlation coefficient and the square correlation coefficient have values approaching 1 as the model accuracy improves.

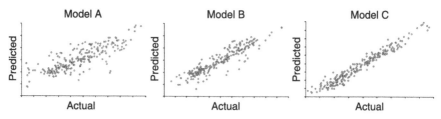

Figure 4.5   Scatterplots of predicted values vs actual values for three models

TABLE 4.5 Comparison of Model Accuracy for Three Models

	Model A	Model B	Model C
Mean square error	1.42	0.622	0.176
Mean absolute error	0.874	0.579	0.333
Relative square error	0.342	0.161	0.051
Relative absolute error	0.52	0.346	0.212
Correlation coefficient	0.811	0.916	0.974
Square correlation coefficient	0.658	0.839	0.949

4.1.4 Evaluating Classification Models' Predictive Accuracy

The overall *accuracy* of a classification prediction model can be estimated by comparing the actual values against those predicted, as long as there are a reasonable number of observations in the test set. An accuracy estimate is calculated using the number of correctly classified observations divided by the total number observations. This results in a number between 0 and 1, where values close to 1 indicates a high accuracy level. The *error rate* or *misclassification rate* is calculated using the number of observations incorrectly classified or one minus accuracy. A confusion matrix, or contingency table, is an effective way of viewing the accuracy of a classification model. For example, Fig. 4.6 shows a table illustrating the results of a classification model. The model's response is the categorical variable *cylinders* which can take five values: 3, 4, 5, 6, and 8. The actual values are shown on the *x*-axis and the predicted values are shown on the *y*-axis. In this example, four observations are predicted as 3; however, only three values are correctly predicted. A single value is incorrectly predicted as 3, when in fact it is 4.

The total number of correctly classified observations can be determined by summing the counts on the diagonal. In Fig. 4.6, that would include 3, 170, 0, 45, and 103, which equal 321. To calculate the overall accuracy, the correct 321

Actual (cylinders)

Predicted (cylinders)		3	4	5	6	8	Totals
	3	3	1	0	0	0	4
	4	1	170	1	32	0	204
	5	0	1	0	0	0	1
	6	0	1	0	45	0	46
	8	0	26	2	6	103	137
	Totals	4	199	3	83	103	392

Figure 4.6   Contingency table showing predicted values against actual values

observations should be divided by the 392 total number of observations, which equals 0.82. One minus this accuracy level is the error rate, or 0.18. Good classification models have high values along the diagonals in the contingency table.

4.1.5 Evaluating Binary Models' Predictive Accuracy

In many situations, prediction models are built for a binary response variable. For example, a model may be built to predict whether an insurance application is fraudulent or not. The ability to predict a fraudulent case may be more important than predicting a nonfraudulent case, so it makes sense to look at the model results in more detail. In this situation, models that minimize false negatives should be selected over those that maximize accuracy.

In the following section, the results from prediction models with a binary response are assessed in greater detail. Counts for the following four properties are initially required:

- *True positive (TP)*: The number of observations predicted to be true (1) that are in fact true (1).

- *True negative (TN)*: The number of observations predicted to be false (0) that are in fact false (0).

- *False positive (FP)*: The number of observations that are incorrectly predicted to be positive (1), but which are in fact negative (0).

- *False negative (FN)*: The number of observations that are incorrectly predicted to be negative (0), but which are in fact positive (1).

These four alternatives are illustrated in the contingency table, or confusion matrix, shown in Table 4.6.

The following values can be calculated to assess the quality of a binary classification prediction model:

- *Accuracy*: The overall accuracy of the model can be calculated based on the number of correctly classified examples divided by the total number of observations,

$$\frac{TP + TN}{TP + FP + FN + TN}$$

TABLE 4.6 Contingency Table Showing the Four Possible Situations

		Actual response	
		Positive (1)	Negative (0)
Prediction	Positive (1)	TP	FP
	Negative (0)	FN	TN

- *Error rate*: The error rate, or misclassification rate, is 1 minus the accuracy value,

$$1 - \frac{TP + TN}{TP + FP + FN + TN}$$

- *Sensitivity*: This is the *true positive rate*, also referred to as the *hit rate*, or *recall*. It is calculated using the number of observations identified as true positives, divided by the actual number of positive observations (TP + FN),

$$\frac{TP}{TP + FN}$$

- *Specificity*: This is the number of negative observations that are correctly predicted to be negative, or the *true negative rate*. It is calculated using the number of correctly predicted negative observations, divided by the total number of actual negative observations (TN + FP),

$$\frac{TN}{TN + FP}$$

- *False positive rate*: This value is the same as 1 minus the sensitivity and is calculated using the number of incorrectly predicted negative observations divided by the actual number of negative observations (FP + TN),

$$\frac{FP}{FP + TN}$$

- *Positive predictive value*: This value is also called *precision*, and it is the number of correctly predicted positive observations divided by the total number of predicted positive observations (TP + FP),

$$\frac{TP}{TP + FP}$$

- *Negative predictive value*: This value is the total number of correctly predicted negative observations divided by the number of negative predictions (TN + FN),

$$\frac{TN}{TN + FN}$$

- *False discovery rate*: This value is the number of incorrectly predicted positive observations divided by the number observations predicted positive (FP + TP),

$$\frac{FP}{FP + TP}$$

	Model A Actual				Model B Actual				Model C Actual		
Predicted		**1**	**0**	Totals	Predicted		**1**	**0**	Totals	Predicted		**1**	**0**	Totals

Model A — Actual:

Predicted	1	0	Totals
1	79	28	107
0	72	213	285
Totals	151	241	392

Model B — Actual:

Predicted	1	0	Totals
1	140	38	178
0	11	203	214
Totals	151	241	392

Model C — Actual:

Predicted	1	0	Totals
1	129	18	147
0	22	223	245
Totals	151	241	392

Figure 4.7   Summary of three different models

TABLE 4.7   Comparison of Different Metrics Across Three Models

	Model A	Model B	Model C
Accuracy	0.75	0.88	0.90
Error	0.26	0.13	0.10
Sensitivity	0.52	0.93	0.86
Specificity	0.88	0.84	0.93
False positive rate	0.12	0.16	0.07
Positive predictive value	0.74	0.79	0.88
Negative predictive value	0.75	0.95	0.91
False discovery rate	0.26	0.21	0.12

Figure 4.7 shows the results from three binary classification models: models A, B, and C. These models show the number of correctly, as well as incorrectly, classified observations, including false positives and false negatives.

Table 4.7 presents an assessment of the three models using the metrics detailed in this section. The overall accuracy and error rate of the models are summarized in the accuracy and error metric. In general, model C is most accurate, followed by model B, and then model A. The metrics also assess how well the models specifically predict positives, with model B performing the best based on the sensitivity score. Model C has the highest specificity score, indicating that this model is the best of the three at predicting negatives.

These different metrics are used in different situations, depending on the goal of the specific project.

4.1.6   ROC Charts

A receiver operating characteristics, or ROC, curve provides an assessment of one or more binary classification models. This chart plots the true positive rate or sensitivity on the y-axis and the false positive rate or 1 minus specificity on the x-axis. Usually a diagonal line is plotted as a baseline, that is, where a random prediction would lie. For classification models that generate a single value, a single point can be plotted on the chart. A point above the diagonal line indicates a degree of accuracy that is better than a random prediction. Conversely, a point below the line indicates that the prediction is worse than a random prediction. The closer the point is to the upper top left point in the chart, the better the prediction.

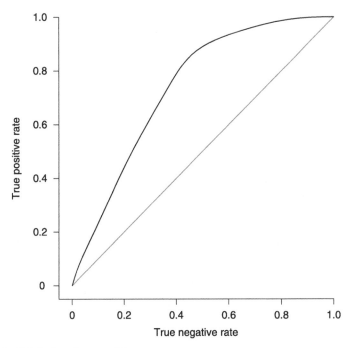

Figure 4.8   ROC chart for a model

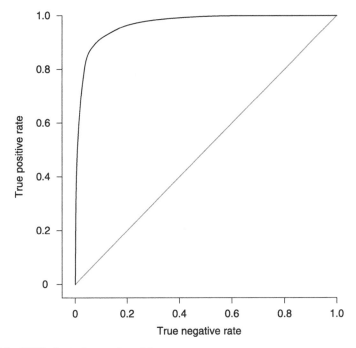

Figure 4.9   ROC chart of a good model

When a classification model generates a numeric value, such as a probability, then a classification can be made by specifying a cutoff threshold. Those numeric predictions above the cutoff are predicted positive, and those below are predicted to be negative. By building multiple models using different threshold cut-offs, a curve can be generated. For example, Fig. 4.8 presents an ROC curve for a model, and Fig. 4.9 shows an ROC curve for a model with a higher level of performance. The area under the curve (AUC) can be used to assess the model's accuracy.

4.1.7 Lift Chart

Many predictive analytics applications require the prediction of a binary response variable. For example, a direct mailing company may wish to predict which households will respond to a specific direct mailing campaign. Those that respond correspond to the positive outcome and those that do not respond correspond to the negative outcome. A predictive model can be built to generate the probability that

TABLE 4.8 Ordered Table of Cumulative Percentages of Observations and Positives

Actual	Prediction probability	Cumulative percent of all observations	Cumulative percentage of positives
1	1	0.3%	0.4%
1	1	0.5%	0.9%
1	1	0.8%	1.3%
1	1	1.0%	1.7%
1	1	1.3%	2.2%
...	...	...	...
1	0.997	25.0%	42.2%
1	0.997	25.3%	42.7%
1	0.997	25.5%	43.1%
1	0.997	25.8%	43.5%
1	0.997	26.0%	44.0%
...	...	...	...
1	0.839	50.0%	81.5%
1	0.837	50.3%	81.9%
1	0.836	50.5%	82.3%
1	0.835	50.8%	82.8%
1	0.834	51.0%	83.2%
...	...	...	...
0	0.0507	75.0%	99.6%
0	0.0499	75.3%	99.6%
0	0.0495	75.5%	99.6%
0	0.0494	75.8%	99.6%
0	0.0478	76.0%	99.6%
...	...	...	...

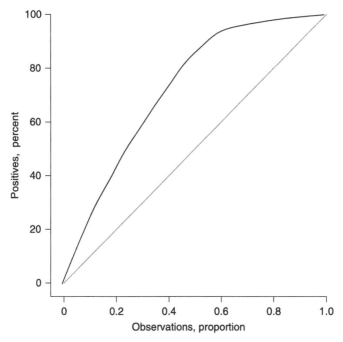

Figure 4.10   Lift chart

a customer will respond. This number would allow the direct mailing company to prioritize customers from a long list of potential households.

A *lift chart* can help to understand how to use a particular model. The chart is based on the original data along with the probability of a positive event generated from a model built from historical data. If this table is ordered according to this probability, those observations with the greatest likelihood of being positive will be at the top of the list. For example, Table 4.8 shows part of a table where a predictive model has been generated for a binary response variable. The actual response is in the column *actual*, and the predicted probability that an observation is true is in the column *probability*. The values in *cumulative percentage of observations* and *cumulative percentage of positives* have been calculated. These last two columns are plotted as a lift chart, as shown in Fig. 4.10. A diagonal line has been included, representing a random outcome.

The lift chart in Fig. 4.10 shows that using the top 50% of the ranked observations will result in approximately 80% of the total positives. For example, if this data represents the direct mailing example, then targeting only the top-ranked 50% of households will result in reaching most of those who will respond. The lift can be calculated at any point, using the target response and dividing it by the average response. In this example, at 50%, the target response is 80%, with an average response at 50%, which gives a lift of 1.6, which is 1.6 times better than not using the model.

4.2 PRINCIPAL COMPONENT ANALYSIS

4.2.1 Overview

Often, data mining projects involve a data set with a large number of continuous variables. If a data set has too many variables, it can be difficult to understand. In addition, the use of all variables in any analysis may introduce a host of logistical and accuracy problems. For example, using a large number of variables may increase the time to compute a particular model beyond an acceptable threshold. Using too many independent variables in a particular model may also impair the accuracy or reliability of a model by overfitting the data. There may even be logistical problems in collecting the values of many variables when the model is deployed.

Principal component analysis provides a method for understanding the meaning of a data set by extracting a smaller series of important components that account for the variability in the data. Each of these factors or *principal components* considers a subset of the variables to be important. For example, a data set containing information on home loans may contain a host of information about an individual, such as salary information, current home price, credit card debt, credit score, and so on. These variables could then be analyzed using principal component analysis. The analysis may group the variables in a number of different ways. For example, variables such as current home price and salary information may be grouped together in a larger group indicating "wealth indicators," whereas credit card debt and credit score may be grouped together as "credit rating" factors.

The use of principal component analysis offers the following benefits:

- *Data set insight*: The process of generating and interpreting the results of a principal component analysis can play an important role in becoming familiar with a data set, and even questioning assumptions about the data. This process may help uncover major factors underlying the data.

- *Reducing the number of variables in the model*: Identifying a smaller set of variables is often helpful, and one approach is to select variables from each important principal component. Alternatively, new variables for each of the important principal components can be generated from the original variables and used as independent variables directly in any modeling exercise.

4.2.2 Principal Components

Each principal component represents a weighted combination of the observed variables. Although all variables in a dataset are combined in a specific principal component, the weights reflect the relative importance of each variable within each principal component. For example, Fig. 4.11 displays a series of weights for five principal components (PC1–PC5) extracted from a dataset of five variables [*age* (*years*), *weight* (*lbs*), *height* (*inches*), *abdomen* (*cm*), *ankle* (*cm*)]. Each *weight*, or *loading*, within a principal component reflects the relative importance of the variable, and these values fall within the range of −1 to +1. For example, in the second principal component, PC2, *age* (*years*) has a strong negative score of −0.781 whereas *weight* (*lbs*) is given a score close to 0.

Principal components	PC1	PC2	PC3	PC4	PC5
Age, year	−0.0186	−0.781	−0.542	0.291	0.105
Weight, lb	0.599	−0.0858	0.0951	−0.28	0.739
Height, in	0.374	0.464	−0.774	−0.104	−0.189
Abdomen, cm	0.521	−0.386	0.223	−0.361	−0.632
Ankle, cm	0.48	0.134	0.221	0.834	−0.0853

Figure 4.11   Five principal components

Principal component analysis produces the same number of components as variables. However, each principal component accounts for a different amount of the variation in the data set. In fact, only a small number of principal components usually account for the majority of the variation in the data. The first principal component accounts for the most variation in the data. The second principal component accounts for the second highest amount of variation in the data, and so on.

Principal component analysis attempts to identify components that are independent of one another; that is, they are not correlated. The first principal component accounts for the largest amount of variation in the data. The second principal component is not correlated to the first; that is, it is *orthogonal* to the first principal component as well as accounting for the second largest remaining variation in the data. The other principal components are generated using the same criteria.

4.2.3   Generating Principal Components

Like most data analysis exercises, principal component analysis starts with a data table comprising a series of observations. Each observation is characterized by a number of variables. The first step in generating the principal components is to construct either a correlation matrix or a covariance matrix. If a covariance matrix is used, the original data may need to be normalized to ensure all variables are on a consistent range. If a correlation matrix is used, this matrix is generated by computing a correlation coefficient (r) for each pair of variables (see Section 3.2.5). For example, Fig. 4.12 shows a correlation matrix formed from a series of 13 variables: *age* (*year*), *weight* (*lbs*), and so on. The variable *age* (*years*) is correlated with each other variable, shown in the first row and first column. For example, the correlation coefficient between *age* (*years*) and *weight* (*lbs*) is −0.0125.

	Age, year	Weight, lb	Height, in	Neck, cm	Chest, cm	Abdomen, cm	Hip, cm	Thigh, cm	Knee, cm	Ankle, cm	Biceps, cm	Forearm, cm	Wrist, cm
Age, year	1	−0.0161	−0.246	0.119	0.182	0.243	−0.0581	−0.216	0.0172	−0.11	−0.0441	−0.0851	0.218
Weight, lb	−0.161	1	0.513	0.81	0.891	0.874	0.933	0.852	0.843	0.581	0.785	0.683	0.725
Height, in	−0.246	0.513	1	0.325	0.224	0.187	0.397	0.35	0.513	0.395	0.319	0.322	0.397
Neck, cm	0.119	0.81	0.325	1	0.769	0.728	0.708	0.669	0.648	0.434	0.709	0.661	0.731
Chest, cm	0.182	0.891	0.224	0.769	1	0.91	0.825	0.708	0.698	0.447	0.707	0.599	0.644
Abdomen, cm	0.243	0.874	0.187	0.728	0.91	1	0.861	0.737	0.71	0.407	0.656	0.53	0.602
Hip, cm	−0.0581	0.933	0.397	0.708	0.825	0.861	1	0.881	0.809	0.521	0.722	0.603	0.626
Thigh, cm	−0.216	0.852	0.35	0.669	0.708	0.737	0.881	1	0.777	0.504	0.744	0.604	0.544
Knee, cm	0.0172	0.843	0.513	0.648	0.698	0.71	0.809	0.777	1	0.585	0.654	0.579	0.656
Ankle, cm	−0.11	0.581	0.395	0.434	0.447	0.407	0.521	0.504	0.585	1	0.449	0.429	0.545
Biceps, cm	−0.0441	0.785	0.319	0.709	0.707	0.656	0.722	744	0.654	0.449	1	0.701	0.614
Forearm, cm	−0.0851	0.683	0.322	0.661	0.599	0.53	0.603	0.604	0.579	0.429	0.701	1	0.598
Wrist, cm	0.218	0.725	0.397	0.731	0.644	0.602	0.626	0.544	0.656	0.545	0.814	0.598	1

Figure 4.12   Correlation matrix

Variance explained:

Principal components	PC1	PC2	PC3	PC4	PC5	PC6	PC7	PC8	PC9	PC10	PC11	PC12	PC13
Eigenvalues	8.05	1.46	0.889	0.661	0.58	0.318	0.284	0.262	0.199	0.143	0.0773	0.0603	0.0181
Percentage	61.9%	11.2%	6.84%	5.08%	4.46%	2.45%	2.18%	2.02%	1.53%	1.1%	0.595%	0.464%	0.139%

Loadings:

	PC1	PC2	PC3	PC4	PC5	PC6	PC7	PC8	PC9	PC10	PC11	PC12	PC13
Age, year	0.00574	0.726	-0.427	-0.079	0.073	-0.359	0.033	-0.108	0.14	-0.298	-0.0922	-0.138	-0.0416
Weight, lb	0.345	-0.0213	0.0288	-0.126	0.0993	0.0773	-0.122	-0.073	-0.0883	0.00318	-0.00717	-0.0544	-0.904
Height, in	0.166	-0.468	-0.517	-0.156	0.555	-0.0241	-0.157	-0.241	-0.0201	-0.156	-0.0999	0.0648	0.177
Neck, cm	0.299	0.137	-0.0323	0.273	0.126	0.555	0.0319	-0.0149	0.683	-0.0243	-0.133	0.0648	0.0649
Chest, cm	0.311	0.237	0.144	-0.11	0.0093	0.173	-0.343	-0.206	-0.201	-0.412	-0.483	-0.338	0.257
Abdomen, cm	0.305	0.281	0.179	-0.275	0.029	0.0561	-0.24	-0.0454	-0.123	-0.0597	0.221	0.751	0.162
Hip, cm	0.325	-0.0225	0.188	-0.268	0.0351	-0.0152	-0.0153	0.126	-0.188	-0.312	0.56	-0.529	0.207
Thigh, cm	0.306	-0.144	0.319	-0.151	-0.0454	-0.0976	0.326	0.204	0.142	-0.484	-0.582	0.066	0.0629
Knee, cm	0.306	-0.0789	-0.127	-0.26	0.029	-0.428	0.199	0.372	0.323	0.583	0.0896	0.0236	0.0291
Ankle, cm	0.221	-0.229	-0.369	-0.11	-0.807	0.0075	-0.152	-0.226	0.113	-0.101	0.0127	0.0121	0.0318
Biceps, cm	0.294	-0.0187	0.172	0.332	0.0268	-0.261	0.482	-0.651	-0.103	0.119	0.137	0.044	0.0467
Forearm, cm	0.264	-0.075	0.058	0.663	0.00797	-0.399	-0.491	0.255	-0.0166	-0.121	-0.00141	0.0215	0.0239
Wrist, cm	0.276	0.116	-0.418	0.257	-0.0514	0.322	0.374	0.379	-0.517	0.0492	-0.04	0.0811	0.0404

Figure 4.13   Extracted principal components along with eigenvalues

The next step is to extract the principal components from this correlation matrix (or covariance matrix), along with the amount of variation explained by each principal component. This is achieved by extracting *eigenvectors* and *eigenvalues* from the matrix (Strang, 2006). The eigenvector is a vector of weights or loadings. The eigenvalues represent the amount of variation explained by each factor. The principal components are then sorted according to the amount of variation they account for. Figure 4.13 illustrates a complete principal component analysis for the same 13 variables as used in Fig. 4.12. An eigenvalue is shown for each principal component along with a percentage of the variance explained by each component. The first principal component (PC1) accounts for 61.9% of the variation in the data, the second principal component (PC2) accounts for 11.2% of the variation in the data, and so on. The weights are also shown for each principal component. For the first principal component, a loading of 0.00574 is assigned to the *age* (*years*) variable, a loading of 0.345 is assigned to the *weight* (*lbs*) variable, and so on. The absolute value of each of the variable's loading values within each principal component represents its relative importance.

Additionally, a new variable can be generated from the original variables' values and the weights of the principal component. For each original data point, the mean for the variable must initially be subtracted from the value (mean centered) and then multiplied by the variable's weight. These values are then summed to create a new variable for each selected principal component. Figure 4.14 shows a scatterplot representing a derived score from principal component 1 against a derived score for principal component 2.

4.2.4 Interpretation of Principal Components

Since the objective of principal component analysis is to identify a small number of factors, the first step is to determine the specific number of principal components to use. Figure 4.15 shows a plot of the variance explained by each principal component, usually referred to as a *scree* plot. The first principal component accounts for the majority of the variation. The ideal number of factors is usually the number just prior to where on the graph the tail levels off. Selecting principal components after this point would add little additional information. In this example, the cutoff should be either at PC2 or PC3.

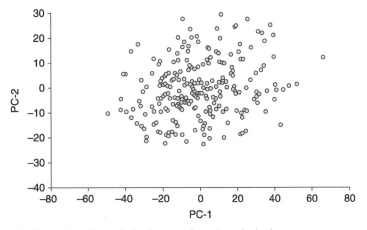

Figure 4.14   Scatterplot of two derived scores from the principal components

Having selected the number of principal components, a process called *rotation of the factors* can help interpret them, if the original analysis is unsatisfactory. This is achieved through a redistribution with these newly rotated principal components now containing loadings towards either $+1$ or -1, with fewer loading values in between. This process also redistributes the amount of variance attributable to each principal component. Methods such as *varimax* (Kaiser, 1958) will perform an optimization on the principal components to accomplish factor rotation. In Fig. 4.16, three

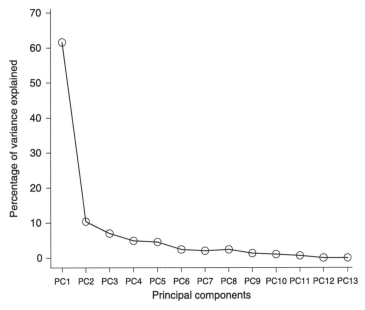

Figure 4.15   Scree plot for the percentage of the variance explained by the principal components

Variance explained:

Principal components	PC1	PC2	PC3
Eigenvalues	8.05	1.46	0.889
Percentage	77.4%	14%	8.55%

Loadings:

Age, year	0.00574	0.726	−0.427
Weight, lb	0.345	−0.0213	0.0288
Height, in	0.166	−0.468	−0.517
Neck, cm	0.299	0.137	−0.0323
Chest, cm	0.311	0.237	0.144
Abdomen, cm	0.305	0.281	0.179
Hip, cm	0.325	−0.0225	0.188
Thigh, cm	0.306	−0.144	0.319
Knee, cm	0.306	−0.0789	−0.127
Ankle, cm	0.221	−0.229	−0.369
Biceps, cm	0.294	−0.0187	0.172
Forearm, cm	0.264	−0.075	0.058
Wrist, cm	0.276	0.116	−0.418

Rotated factors:

Principal components	PC1 (rot)	PC2 (rot)	PC3 (rot)
Age, year	−0.00594	0.84	0.0532
Weight, lb	0.33	−0.0217	−0.105
Height, in	−0.074	−0.097	−0.706
Neck, cm	0.294	0.142	−0.0532
Chest, cm	0.374	0.128	0.132
Abdomen, cm	0.385	0.145	0.185
Hip, cm	0.359	−0.112	0.0239
Thigh, cm	0.36	−0.285	0.0674
Knee, cm	0.237	0.0156	−0.244
Ankle, cm	0.0606	0.0216	−0.483
Biceps, cm	0.325	−0.101	0.0242
Forearm, cm	0.253	−0.0853	−0.0839
Wrist, cm	0.154	0.338	−0.356

Figure 4.16   The original loadings along with the rotated factors

principal components were selected and rotated. In this example, by comparing the original values to the new rotated values, the second principal component, *age* (*years*), is now closer to +1 while the others, with the exception of *wrist* (*cm*), are closer to 0. Once a set of selected and rotated principal components is identified, the final step is to name them, using the weights as a guide to assist the analysis. Section 5.5 provides an example of the use of principal component analysis.

4.3   MULTIPLE LINEAR REGRESSION

4.3.1   Overview

Multiple linear regression analysis is a popular method used in many data mining projects for building models to predict a continuous response variable. This model defines the linear relationship between a series of independent variables and a single response variable. It can be used to generate models, for example, to predict sales from transactional data, or to predict credit scores from information in a person's credit history.

The use of multiple linear regression analysis has a number of advantages, including:

- *Easy to understand*: Multiple linear regression models are easy to understand and interpret, because they are represented as a weighted series of independent variables. They can be effective in predicting new data as well as explaining what variables are influential within the data set.

- *Detect outliers*: In addition to this method's use as a prediction model, it can also help identify outliers, that is, those observations that do not follow a linear trend observed by the other entries.

- *Fast*: The generation of a multiple linear regression equation is fast, and it enables the rapid exploration of alternative variables since multiple models can be quickly built using different combinations of variables to determine an optimal model.

Despite being an effective method for prediction from and explanation of a data set, multiple linear regression analysis has a number of disadvantages, including:

- *Sensitivity to noise and outliers*: Multiple linear regression models are sensitive to noisy data as they try to find a solution that best fits all data, including the outliers. Outliers are erroneous pieces of data that can have especially undesirable consequences as the model tries to fit the potentially erroneous values.

- *Only linear relationships handled*: These models cannot model nonlinear data-sets; however, the calculation of new variables can help in modeling. Transforming the independent and/or the response variables using mathematical transformation such as log, squared, cubed, square root, and so on, can help to incorporate variables with nonlinear relationships.

The simplest form of a linear regression is one containing a single independent variable, also referred to as *simple linear regression*. In this situation, the model can be drawn as a straight line through the data, plotted on a scatterplot. For example, Fig. 4.17 shows a scatterplot of two variables A and B, and a line drawn through them to represent a linear model. This linear model is represented by the formula for the straight line:

$$A = -0.27 + 1.02 \times B$$

Multiple linear regression analysis involves understanding the relationship between more than one independent variable and a single response variable. The analysis does not imply that one variable causes another variable to change; it only recognizes the presence of a relationship. This relationship is difficult to visualize when dealing with more than one or two independent variables. The relationship between the response variable and the independent variables for the entire population is assumed to be a linear equation of the form:

$$y = \beta_0 + \beta_1 x_1 + \beta_2 x_2 + \cdots + \beta_k x_n + \varepsilon$$

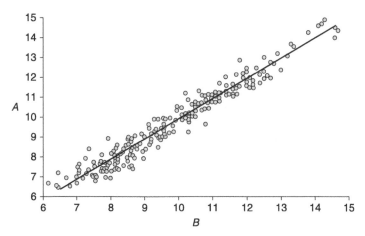

Figure 4.17   Illustration of a simple linear regression model, represented as a straight line

In this equation, y is the response, $\beta_0-\beta_k$ are constant values referred to as *beta coefficients*, x_1-x_n are the input independent variables, and ε is a random error. Since the model will be built from a sample of the population, building a multiple linear regression model will estimate values for the beta coefficients, and the equation generated will look like:

$$\hat{y} = \hat{\beta}_0 + \hat{\beta}_1 x_1 + \hat{\beta}_2 x_2 + \cdots + \hat{\beta}_k x_n$$

In this equation, $\hat{y}$ is the predicted response, x_1-x_n are the independent variables, and $\hat{\beta}_0$ through $\hat{\beta}_k$ are the estimated values of the beta coefficients.

For example, a multiple linear regression equation to predict a credit score, using two variables *LOANS* and *MISSED_PAYMENTS* may look like:

$$CREDIT\_SCORE = 22.5 + 0.5 \times LOANS - 0.8 \times MISSED\_PAYMENTS$$

A number of assumptions must be made when building a multiple linear regression model. These assumptions can be tested once the model has been built, as described later in this chapter. These assumptions are:

- *Linear*: A multiple linear regression will only generate models that describe a linear relationship between the independent variables and the response.
- *Homoscedasticity*: This refers to the assumption that the variation of error terms should be constant with respect to the independent variables; that is, there should be no relationship between the independent variable's variation and the error term.
- *Independence*: The error values should not be a function of any adjacent values, for example, to avoid errors that result from the passage of time.
- *Normally distributed error term*: The frequency distribution of the errors (predicted value minus the actual value) is assumed to follow a normal distribution.

The variables used as independent variables should not be correlated to one another. This situation, known as *multicolinearity*, will cause the models to fail. A scatterplot can be used to check for multicolinearity. Correlations between continuous variables, as well as categorical variables, should be checked. An example of a multicolinearity situation involving continuous variables is a dataset used to predict a health condition. This data set may contain two or more tests that, in fact, measure the same phenomena, and hence only one should be included. A multicolinearity situation involving categorical variables involves a model used to predict the success of a marketing campaign. In this example, the independent variable *color* can take three values: red, green and blue. To use this variable within the model, the *color* variable is transformed into three dummy variables corresponding to each color, as discussed in Chapter 1. This results in the generation of three new variables; however, an observation where blue is 1 always occurs when red and green are 0. Hence, only two variables are really needed to capture all possible scenarios. The inclusion of all three variables would include correlations between the *color* variables and therefore all three should not be used. After a model is built, multicolinearity may be observed through interpreting the beta coefficients to look for, as an example, unexpected positive or negative numbers that do not reflect how the model is expected to behave.

The following sections discuss how a multiple linear regression model is built, and, once built, how its assumptions are tested through an analysis of the errors. Most statistical software packages generate a series of statistics concerning the models built, such as the *standard error* and the *coefficient of multiple determination*, along with metrics to assess the significance of the models and the parameters. Finally, in building a model, alternative combinations of variables should be used to build multiple models, and alternative data transformations (such as product of variables or variables squared) can be considered to improve the quality of the models built.

4.3.2 Generating Models

A multiple linear regression model is an equation describing the linear relationship between a response variable and a series of independent variables. The equation is a weighted sum of all variables, where $\hat{\beta}_1 - \hat{\beta}_k$ correspond to the weights and $x_1 - x_n$ correspond to the independent variables. $\hat{y}$ is the response variable being predicted and $\hat{\beta}_0$ is a constant added to the equation.

$$\hat{y} = \hat{\beta}_0 + \hat{\beta}_1 x_1 + \hat{\beta}_2 x_2 + \cdots + \hat{\beta}_k x_n$$

The multiple linear regression formula can be rewritten using matrix multiplication:

$$y = X\hat{\beta}$$

The response variable, y, is a column vector of values, where $y_1 - y_n$ are the response values for the n observations:

$$y = \begin{pmatrix} y_1 \\ y_2 \\ y_3 \\ \cdots \\ y_n \end{pmatrix}$$

The independent variables, X, are represented as a matrix, where n is the number of observations and k is the number of variables to be used as independent variables. The first column is all 1s and relates to the intercept $\hat{\beta}_0$.

$$X = \begin{pmatrix} 1 & x_{1,1} & x_{1,2} & \cdots & x_{1,k} \\ 1 & x_{2,1} & x_{2,2} & \cdots & x_{2,k} \\ 1 & x_{3,1} & x_{3,2} & \cdots & x_{3,k} \\ \cdots & \cdots & \cdots & \cdots & \cdots \\ 1 & x_{n,1} & x_{n,2} & \cdots & x_{n,k} \end{pmatrix}$$

The β coefficients are also described as a vector, where $\hat{\beta}_0 - \hat{\beta}_n$ are the individual coefficients:

$$\hat{\beta} = \begin{pmatrix} \hat{\beta}_0 \\ \hat{\beta}_1 \\ \hat{\beta}_2 \\ \cdots \\ \hat{\beta}_n \end{pmatrix}$$

To generate a multiple linear regression model, estimates for the β coefficients are derived from the training data. The objective of the process is to identify the best fitting model for the data. A procedure referred to as *least squares* attempts to derive a set of coefficients to minimize the model's error (ε). This error is assessed using the *sum of squares of error* (SSE), such that:

$$\text{SSE} = \sum_{i=1}^{n} \varepsilon_i^2$$

The formula calculated is based on the error (ε) squared. The error is squared so that positive and negative errors do not cancel each other out. The error is calculated from the difference between the predicted value ($\hat{y}_i$) and the actual response value (y_i):

$$\text{SSE} = \sum_{i=1}^{n} (y_i - \hat{y}_i)^2$$

Replacing $\hat{y}_i$ with the equation for the multiple linear regression results in the following:

$$\text{SSE} = \sum_{i=1}^{n} (y_i - \hat{\beta}_0 - \hat{\beta}_1 x_1 - \cdots - \hat{\beta}_k x_k)^2$$

This equation is then solved using calculus, and the details of this calculation are provided in Rencher (2002). The β coefficients can then be calculated using the following matrix formula:

$$\hat{\beta} = (X^T X)^{-1} X^T y$$

In this formula the superscript T represents a transposed matrix and the superscript -1 represents an inverse matrix. This calculation is always performed with a

TABLE 4.9   Table of Data Relating Rental Prices to Their Square Footage and Number of Baths

SQUARE_FEET	NOS_BATHS	RENTAL_PRICE
789	1	770
878	1	880
939	2	930
1100	2	995
1300	3	1115
1371	3	1300
1481	3	1550
750	1	560
850	1	610
2100	3	1775
1719	3	1450
1900	3	1650
1100	3	900
874	1	673
1024	2	785
1082	2	809

computer because of the complexity of the matrix operations. See Appendix A for more details.

In the following example, a multiple linear regression model is built to predict the rental price of apartments in a specific neighborhood. The response variable is *RENTAL_PRICE* and the independent variables used are *SQUARE_FEET* and *NOS_BATHS*. Table 4.9 is a data table to be used to build the model.

The independent variables, *SQUARE_FEET* and *NOS_BATHS*, are converted to a matrix, with the first column all 1s (corresponding to the intercept), the second columns *SQUARE_FEET* and the third column *NOS_BATHS*:

$$
X = \begin{pmatrix} 1 & 789 & 1 \\ 1 & 878 & 1 \\ 1 & 939 & 2 \\ \cdots & \cdots & \cdots \\ 1 & 1082 & 2 \end{pmatrix}
$$

The response variable, *RENTAL_PRICE*, is converted to a vector:

$$
y = \begin{pmatrix} 770 \\ 880 \\ 930 \\ \cdots \\ 809 \end{pmatrix}
$$

The β coefficients of the model are calculated using:

$$\hat{\beta} = (X^T X)^{-1} X^T y$$

$$
\hat{\beta} = \left[\begin{pmatrix} 1 & 789 & 1 \\ 1 & 878 & 1 \\ 1 & 939 & 2 \\ \dots & \dots & \dots \\ 1 & 1082 & 2 \end{pmatrix}^T \begin{pmatrix} 1 & 789 & 1 \\ 1 & 878 & 1 \\ 1 & 939 & 2 \\ \dots & \dots & \dots \\ 1 & 1082 & 2 \end{pmatrix} \right]^{-1} \begin{pmatrix} 1 & 789 & 1 \\ 1 & 878 & 1 \\ 1 & 939 & 2 \\ \dots & \dots & \dots \\ 1 & 1082 & 2 \end{pmatrix}^T \begin{pmatrix} 770 \\ 880 \\ 930 \\ \dots \\ 809 \end{pmatrix}
$$

$$
\hat{\beta} = \begin{pmatrix} -33.3 \\ 0.816 \\ 46.5 \end{pmatrix}
$$

The equation for the model relating the square feet and number of baths for apartments to the rental price is therefore:

$$RENTAL\_PRICE = -33.3 + 0.816 \times SQUARE\_FEET + 46.5 \times NOS\_BATHS$$

4.3.3 Prediction

To make a prediction of the rental price for an apartment with 912 square feet and one bathroom, these values are substituted into the equation, thus resulting in a prediction of 757 for the *RENTAL_PRICE*:

$$RENTAL\_PRICE = -33.3 + 0.816 \times SQUARE\_FEET + 46.5 \times NOS\_BATHS$$
$$RENTAL\_PRICE = -33.3 + 0.816 \times 912 + 46.5 \times 1$$
$$RENTAL\_PRICE = 757$$

The coefficients define the rate at which the model's prediction will change as the independent variables change, when all other independent variables are kept the same. The higher the coefficient, the greater the change. For example, increasing the number of baths in this example to 2, while keeping the square feet the same will increase the rental price by 46.5.

It is usual to make a prediction from data within the same range as the data used to build the model.

4.3.4 Analysis of Residuals

Once a model has been built, a prediction can be computed for each observation by using the actual values for the x-variables within the model and calculating a predicted value for the y variables ($\hat{y}$). For example, in the model previously built for the apartment rental example, a prediction has now been calculated using the regression model and the actual x-variables, *SQUARE_FEET* and *NOS_BATHS*. Table 4.10 shows the predicted values along with the residuals.

TABLE 4.10 Calculations of Predictions and Residuals

RENTAL_PRICE, y	SQUARE_FEET, x_1	NOS_BATHS, x_2	PREDICTION, $\hat{y}$	RESIDUAL, ERR
770	789	1	657	113
880	878	1	729	151
930	939	2	825	105
995	1100	2	957	38
1115	1300	3	1166	−51
1300	1371	3	1224	76
1550	1481	3	1314	236
560	750	1	625	−65
610	850	1	706	−96
1775	2100	3	1819	−44
1450	1719	3	1508	−58
1650	1900	3	1656	−6
900	1100	3	1003	−103
673	874	1	726	−53
785	1024	2	895	−110
809	1082	2	942	−133

For each predicted value, the difference between what the model predicts and the actual value is referred to as the *error* or the *residual*. For example, in Table 4.10, the first observation has an actual value for *RENTAL_PRICE* of 770, and a predicted value of 657. The difference between these two values reflects the error or residual.

$$\text{residual} = 770 - 657 = 113$$

A residual has been calculated for each observation in the data set, as shown in Table 4.10. Since the regression model has been calculated to minimize errors, the sum of all residual values should equal zero. Analysis of residuals is helpful in testing the following underlying assumptions, also mentioned above, of multiple linear regression, that is:

- *Linear*: A multiple linear regression will only generate linear models, and hence understanding whether the relationship is in fact linear is important. This can be seen in Fig. 4.18, where the residual is plotted against the *y*-variable and a clear "U"-shape can be seen, indicating a nonlinear relationship. Plotting the residual against the individual independent variables will help to identify nonlinear relationship that could be rectified with a mathematical transformation, such as a quadratic term.

- *Homoscedasticity*: This refers to the assumption that the variation of error terms should be constant with respect to the independent variables. This can be tested by plotting, for example, the predicted variable against the residual. If there is a trend indicating a nonconstant variance, as shown in Fig. 4.19, then the underlying assumption of homoscedasticity is not valid.

- *Independence*: The error values should not be a function of any adjacent values, and this can easily be tested by plotting the residual values against

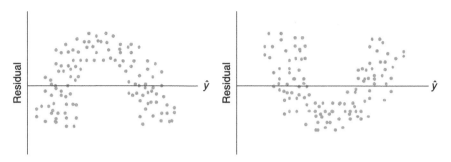

Figure 4.18   Nonlinear relationship shown by plotting the predicted variable against the residual

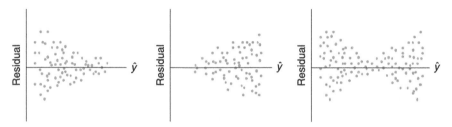

Figure 4.19   A model that violates the homoscedasticity assumption, as seen by plotting the residual against the prediction

the order in which the values were collected. In Fig. 4.20, the residual values have been plotted against the order the observations were taken, and a clear trend is discernable in both graphs, indicating that the assumption of independence is violated. This error may have been introduced as a result of the measurements being taken over time.

- *Normally distributed error term*: Examining the frequency distribution of the residuals, for example, using a frequency histogram or a q–q plot (as discussed in Chapter 2), is helpful in assessing whether the normal distribution assumption is violated. This assumption is usually required to enable computations of confidence intervals, which are not required in many data mining applications.

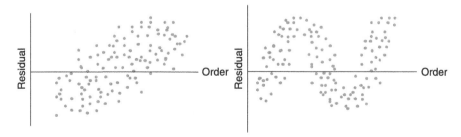

Figure 4.20   Plot of residual against the order the observations were collected, indicating a violation of the assumption of independence

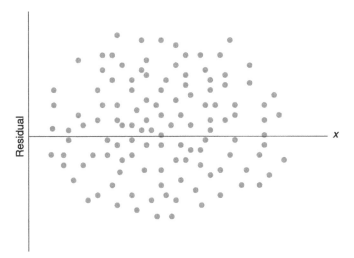

Figure 4.21   Residuals plotted against the x-variable with no discernable trend

There should be no discernable trend in the residual plot of any of the independent variables against the errors, as shown in Fig. 4.21.

The analysis of residuals is helpful in determining whether there are any clear violations in the assumptions. An analysis of the residual plots can also be helpful in identifying observations that do not fit the model, that is, those observations with unusually high positive or negative residual values. These outlier observations may be attributable to errors in the data and should be examined in more detail to determine whether to remove them.

4.3.5   Standard Error

An evaluation of the residuals in the model can help in understanding whether the model is violating any assumptions. An overall assessment of the model error is computed using the SSE. The following formula, as described earlier, is used to calculate the SSE:

$$\text{SSE} = \sum_{i=1}^{n}(y_i - \hat{y}_i)^2$$

Using Table 4.10, the SSE can be calculated as:

$$\text{SSE} = \sum_{i=1}^{n}(y_i - \hat{y}_i)^2$$

$$\text{SSE} = (770 - 113)^2 + (880 - 729)^2 + \cdots + (809 - 942)^2$$

$$\text{SSE} = 174{,}096$$

The distribution or spread of the residual values can also be useful in assessing the model. This is achieved by calculating the standard deviation of the residual or the *standard error of the estimate* S_e. Assuming that the error terms are normally distributed, approximately 68% of errors will be within one standard deviation, and approximately 95% of errors will be within two standard deviations. The formula for the standard error of the estimate S_e is:

$$S_e = \sqrt{\frac{SSE}{n - k - 1}}$$

where SSE is the sum of the squares of errors, n is the number of observations, and k is the number of independent variables.

In the example model, where SSE is 174,096, the number of observations is 16 (n), and the number of independent variables or k is 2, S_e is:

$$S_e = \sqrt{\frac{SSE}{n - k - 1}}$$

$$S_e = \sqrt{\frac{174{,}096}{16 - 2 - 1}}$$

$$S_e = 112$$

This value can help in assessing whether the model is sufficiently accurate. Assuming a normal distribution, approximately 68% of errors should be within one standard deviation or ±112 and approximately 95% of errors should be within two standard deviations, that is, ±224.

4.3.6 Coefficient of Multiple Determination

Most statistical software packages that perform a multiple linear regression analysis also calculate the coefficient of multiple determination, or R^2. This coefficient is used to assess how much of the variation in the response is explained by the model. It is determined using the difference between the variance in the data about a naive model, where the mean response is used as the model, against the variance attributable to the fitted model. The value for R^2 varies between 0 and 1, with a high value indicating that a significant portion of the variance in the response is explained by the model. The formula to calculate R^2 is based on the SSE and the *total sum of square* (SST), or error about a naive model. To calculate SST, using $\bar{y}$ as the mean value for y:

$$SST = \sum_{i=1}^{n} (y_i - \bar{y})^2$$

The sum of squares of error has been previously discussed and is calculated using:

$$SSE = \sum_{i=1}^{n} (y_i - \hat{y}_i)^2$$

The coefficient of determination is therefore calculated using the difference in the variation explained by the fitted model and the naive model (explained variability) as a proportion of the total error:

$$R^2 = \frac{\text{SST} - \text{SSE}}{\text{SST}}$$

In the apartment rental example, using a mean *RENTAL_PRICE* value of 1047 ($\bar{y}$), the SST can be calculated as:

$$\text{SST} = \sum_{i=1}^{n} (y_i - \bar{y})^2$$

$$\text{SST} = (770 - 1047)^2 + (880 - 1047)^2 + \cdots + (809 - 1047)^2$$

$$\text{SST} = 2{,}213{,}566$$

Using Table 4.10, SSE can be calculated as:

$$\text{SSE} = \sum_{i=1}^{n} (y_i - \hat{y}_i)^2$$

$$\text{SSE} = (770 - 113)^2 + (880 - 729)^2 + \cdots + (809 - 942)^2$$

$$\text{SSE} = 174{,}096$$

In the apartment example, the percentage of variation in the response as explained by the model is therefore:

$$R^2 = \frac{\text{SST} - \text{SSE}}{\text{SST}}$$

$$R^2 = \frac{2{,}213{,}566 - 174{,}096}{2{,}213{,}566}$$

$$R^2 = 0.92$$

An increasing number of independent variables result in an R^2 value that is overestimated. An adjusted R^2 or R^2_{adj} is usually calculated to more accurately reflect the number of independent variables, as well as the number of observations:

$$R^2_{\text{adj}} = 1 - \left(\frac{n-1}{n-k-1} \right) (1 - R^2)$$

In the apartment rental example, there are 16 observations (n) and two independent variables (k), and the following value of R^2_{adj} is calculated:

$$R^2_{adj} = 1 - \left(\frac{n-1}{n-k-1}\right)(1 - R^2)$$

$$R^2_{adj} = 1 - \left(\frac{16-1}{16-2-1}\right)(1 - 0.92)$$

$$R^2_{adj} = 0.908$$

Usually, R^2_{adj} values are slightly less than R^2 values.

4.3.7   Testing the Model Significance

Assessing the significance of the relationship between the independent variables and the response is an important step. An F-test is most often used, based on the following hypothesis:

$$H_0: \beta_1 = \beta_2 = \cdots = \beta_K = 0$$
$$H_a: \text{At least one of the coefficients is not equal to } 0$$

The null hypothesis states that there is no linear relationship between the response and the independent variables. If the null hypothesis is rejected, it is determined that there is a significant relationship. An F-test is performed using the mean square regression (MSR) and the mean square error (MSE). The formula for MSR is:

$$MSR = \frac{SSR}{k}$$

The formula for MSE is:

$$MSE = \frac{SSE}{n-k-1}$$

The F-test is calculated using the formula:

$$F = \frac{MSR}{MSE}$$

The *regression sum of squares* (SSR) is calculated using the following formula:

$$SSR = \sum_{i=1}^{n}(\hat{y}_i - \bar{y}_i)^2$$

For the apartment example, it is calculated as 2,040,244. Using this value, the MSR is:

$$MSR = \frac{SSR}{k}$$

$$MSR = \frac{2,040,244}{2}$$

$$MSR = 1,020,122$$

Using the previously calculated value for the SSE, the value for MSE is:

$$MSE = \frac{SSE}{n - k - 1}$$

$$MSE = \frac{174,096}{16 - 2 - 1}$$

$$MSE = 76$$

An F-value of 76 is calculated and this number is compared to the critical F-value in order to determine whether the null hypothesis is rejected. The critical value is based on the level of significance (α), the degrees of freedom of the regression (k), and the degrees of freedom of the error ($n - k - 1$). Assuming a level of significance of 0.01, the critical value for $F_{0.01,2,14}$ is 6.51, using a standard F-distribution table (see Myatt, 2007). Since the computed F-value is greater than the critical value, the null hypothesis is rejected. A p-value is usually computed in most statistical software packages and can also be used to make this assessment.

In addition to assessing the overall model, each individual coefficient can be assessed. A t-test is usually performed, based on the following hypothesis:

$$H_0: \beta_j = 0$$
$$H_a: \beta_j \neq 0$$

The null hypothesis states that the coefficient is not significant. In the apartment rental example, the independent variable *SQUARE_FEET* has a calculated t-value of 6.53 and a p-value of almost 0, indicating the significance of this variable; however, the *NOS_BATHS* variable has a t-value of 0.803 with a p-value of 0.44, indicating that this variable is less significant within the model for some reason.

4.3.8   Selecting and Transforming Variables

Calculating which variable combinations result in the best model is often determined by evaluating different models built with different combinations of independent variables. Each model may be checked against an indication of the quality of the model, such as R^2_{adj}. An *exhaustive search* of all possible variable combinations is one approach to identifying the optimal set of independent variables. This approach, however, can be time-consuming and methods such as *forward selection, backward selection*, and *stepwise selection* provide faster methods for identifying independent variable combinations. These methods add and remove variables based on different rules, and they will identify solutions more quickly, with the risk of overlooking the best solution. The forward selection method adds independent variables one at a time, building on those additions that result in an increase in the performance of the model. The backwards selection method starts with all independent variables, and sequentially removes variables that do not contribute to the model performance. Finally the stepwise method can proceed in the forward or backward direction and assesses the contribution of the variables at each step. The Further Reading section of this chapter points to additional material on these approaches.

The following example illustrates the process of building multiple models with different sets of independent variables. In this example, a series of models were built to predict *percentage body mass* (the response variable), using up to four independent variables: *weight, chest, abdomen*, and *hip*. The exhaustive search method is used to generate models using all combinations of the four independent variables, and an adjusted R^2 is calculated for each model generated. It can be seen from Table 4.11 that a model built from two independent variables, *weight* and *abdomen*, yields a model with the highest adjusted R^2 value of 0.716.

When a model would violate one of the underlying assumptions, various mathematical transformations could be applied to either the independent variables or response variables, or both. Transformations such as the natural log, polynomials, reciprocals, and square roots can aid in building multiple linear regression models.

TABLE 4.11 Building Different Models with All Combinations of Independent Variables

Variable 1	Variable 2	Variable 3	Variable 4	R^2_{adj}
Weight				0.371
Chest				0.492
Abdomen				0.659
Hip				0.382
Weight	Chest			0.492
Weight	Abdomen			0.716
Weight	Hip			0.386
Chest	Abdomen			0.668
Chest	Hip			0.494
Abdomen	Hip			0.693
Weight	Chest	Abdomen		0.714
Weight	Chest	Hip		0.515
Weight	Abdomen	Hip		0.715
Chest	Abdomen	Hip		0.694
Weight	Chest	Abdomen	Hip	0.713

4.4   DISCRIMINANT ANALYSIS

4.4.1   Overview

Discriminant analysis is used to make predictions when the response variable is categorical (a classification model). For example, an insurance company may want to predict high or low risk customers, or a marketing department may want to predict whether a current customer will or will not buy a particular product. Discriminant analysis models will classify observations based on a series of independent variables. Figure 4.22 illustrates its use with a data set. Two variables are used to show the distribution of the data. The light circles belong to group A, and the dark circles represent observations in group B. The objective of the modeling exercise is to classify observations into either group A or group B. A straight line has been drawn to illustrate how, on one side of the line, the observations are assigned to group A and on the other they are assigned to group B. Discriminant analysis attempts to find a straight line that separates the two classes, and provides a method for assigning new observations to one of the classes.

The analysis becomes more complex in situations when the data set contains more independent variables, as well as when the response variable has more than two possible outcomes. In these situations, discriminant analysis attempts to identify hyperplanes that separate these multiple groups.

Discriminant analysis is a simple statistical technique that can be used to identify important variables that characterize differences between groups, as well as to build classification models. It is a useful classification method, especially for smaller data sets. The following is a summary of the key assumptions associated with using discriminant analysis:

- *Multivariate normal distribution*: The variables should have a normal distribution within the classes.

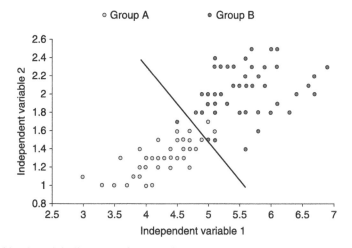

Figure 4.22   A straight line separating two classes (group A and group B)

- *Similar group covariance matrices*: In addition, correlations between and variances of the independent variables in each group used in the model should be similar.

Discriminant analysis is sensitive to outliers and will not operate well in situations where the size of one or more of the groups is small.

4.4.2   Discriminant Function

The method relies on the calculation of a *discriminant function* for each group. There will be k functions generated based on the number of unique categories the response variable can take. If the response variable is *color* with possible values red, green, and blue, then k will be 3. Predictions are made by calculating a score using each group's discriminant function. An observation is predicted to be a member of the group with the highest discriminant function score.

Similar to multiple linear regression, linear discriminant analysis attempts to identify a solution that minimizes the overall error. By making the assumptions described earlier, that is, a normal distribution, with similar group covariance matrices, the following formula can be used to estimate a linear discriminant function:

$$f_k = x^T \hat{S}^{-1} \hat{\mu}_k - \frac{1}{2} \hat{\mu}_k^T \hat{S}^{-1} \hat{\mu}_k + \log(\hat{p}_k)$$

In this formula, x is a vector of values for a single observation, $\hat{S}$ is an estimate of the covariance matrix, $\hat{\mu}_k$ is a vector of mean values for the variables corresponding to group k, and $\hat{p}_k$ is an estimate of the prior probability. The superscript -1 represents an inverse matrix and the superscript T represents a transposed matrix. More information on how this formula was derived can be found in Hastie (2003) and Rencher (2002).

One approach to calculating $\hat{S}$, the estimate of the covariance, is to calculate the covariance matrix (described in Section 3.3.5) for each of the groups and then combine the individual matrices into a single pooled covariance matrix.

$\hat{p}_k$ represents the prior probability and can be estimated using N_k, which is the number of observations in category k:

$$\hat{p}_k = \frac{N_k}{N}$$

4.4.3   Discriminant Analysis Example

As an example, a data set of wines that includes a number of their chemical properties can demonstrate these ideas (http://archive.ics.uci.edu/ml/datasets/Wine). Each wine has an entry for the variable *alcohol*, which can take three values: "1," "2," and "3," that relate to the wine's region. A discriminant analysis model will be built to predict this response. The wines are described using a number of independent variables: (1) *malic acid*, (2) *alkalinity of ash*, (3) *nonflavanoids*, and (4) *proline*. There are 198 observations and Table 4.12 presents a number of example observations, which are further summarized in Figs. 4.23–4.27. In Figs.

TABLE 4.12 Data Table Illustrating the Alcohol Classification

Alcohol	Malic acid	Alkalinity of ash	Nonflavanoids	Proline
1	14.23	2.43	3.06	1065
1	13.2	2.14	2.76	1050
1	13.16	2.67	3.24	1185
1	14.37	2.5	3.49	1480
1	13.24	2.87	2.69	735
1	14.2	2.45	3.39	1450
1	14.39	2.45	2.52	1290
1	14.06	2.61	2.51	1295
1	14.83	2.17	2.98	1045

4.23–4.26, the frequency distribution of the four independent variables is presented, with highlighting indicating the three alcohol classes. Similarly, Fig. 4.27 presents a scatterplot matrix of the four independent variables, with the three alcohol classes highlighted.

The three classes are initially summarized as shown in Table 4.13, where a count of the number of observations in each class is presented along with the mean for the four independent variables, corresponding to each class.

One approach to estimating the covariance matrix, detailed in Section 3.3.5, needed for the discriminant function is to calculate a covariance matrix for each group and then to pool the values. This results in the following covariance matrix:

$$\hat{S} = \begin{bmatrix} 0.67 & 0.048 & 0.19 & 166 \\ 0.048 & 0.076 & 0.032 & 19.5 \\ 0.19 & 0.032 & 1.01 & 157.4 \\ 166.2 & 19.5 & 157.4 & 100{,}249 \end{bmatrix}$$

This is inverted, resulting in the following matrix:

$$\hat{S}^{-1} = \begin{bmatrix} 2.62 & -0.56 & 0.23 & -0.0046 \\ -0.56 & 13.96 & -0.070 & -0.0017 \\ 0.23 & -0.070 & 1.33 & -0.0025 \\ -0.0046 & -0.0017 & -0.0025 & 0.000022 \end{bmatrix}$$

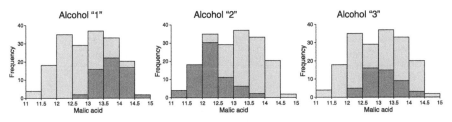

Figure 4.23   Three alcohol groups highlighted on the variable malic acid

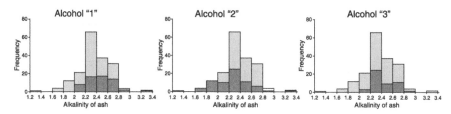

Figure 4.24   Three alcohol groups highlighted on the variable alkalinity of ash

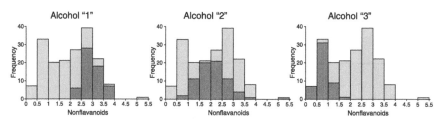

Figure 4.25   Three alcohol groups highlighted on the nonflavanoids variable

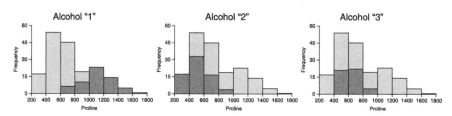

Figure 4.26   Three alcohol groups highlighted on the proline variable

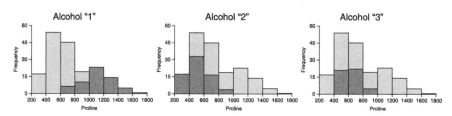

Figure 4.27   Three scatterplot matrices showing the three alcohol classes highlighted over a matrix of the four independent variables

TABLE 4.13   Summary Table of Different Wine Groups

Group name	Count	Mean malic acid	Mean alkalinity of ash	Mean nonflavanoids	Mean proline
"1"	59	13.7	2.46	2.98	1116
"2"	71	12.3	2.25	2.08	520
"3"	48	13.2	2.44	0.78	630

TABLE 4.14 An Observation to be Used in the Discriminant Analysis Model

Malic acid	Alkalinity of ash	Nonflavanoids	Proline
14.23	2.43	3.06	1065

The value p_k is the prior probability and can be estimated using N_k, which is the number of observations in category k:

$$p_k = \frac{N_k}{N}$$

For example, to calculate the prior probability for group 1:

$$p_1 = \frac{59}{198} = 0.33$$

Using this information, a discriminant function score can be calculated for each observation, for each class. In this example, three scores are calculated for each of the response categories, and the values for the independent variables are shown in Table 4.14.

The "1" group function is calculated using the following formula:

$$f_1 = x^T \hat{S}^{-1} \hat{\mu}_1 - \frac{1}{2} \hat{\mu}_1^T \hat{S}^{-1} \hat{\mu}_1 + \log(p_1)$$

To calculate the first part of the equation, $x^T \hat{S}^{-1} \hat{\mu}_1$:

$$x^T \hat{S}^{-1} \hat{\mu}_1 = \begin{bmatrix} 14.23 & 2.43 & 3.06 & 1065 \end{bmatrix}$$

$$\begin{bmatrix} 2.62 & -0.56 & 0.23 & -0.0046 \\ -0.56 & 13.96 & -0.070 & -0.0017 \\ 0.23 & -0.070 & 1.33 & -0.0025 \\ -0.0046 & -0.0017 & -0.0025 & 0.000022 \end{bmatrix} \begin{bmatrix} 13.7 \\ 2.46 \\ 2.98 \\ 1116 \end{bmatrix}$$

$$x^T \hat{S}^{-1} \hat{\mu}_1 = 449$$

To calculate the second part of the equation, $\frac{1}{2} \hat{\mu}_1^T \hat{S}^{-1} \hat{\mu}_1$:

$$\frac{1}{2} \hat{\mu}_1^T \hat{S}^{-1} \hat{\mu}_1 = \frac{1}{2} \begin{bmatrix} 13.7 & 2.46 & 2.98 & 1116 \end{bmatrix}$$

$$\begin{bmatrix} 2.62 & -0.56 & 0.23 & -0.0046 \\ -0.56 & 13.96 & -0.070 & -0.0017 \\ 0.23 & -0.070 & 1.33 & -0.0025 \\ -0.0046 & -0.0017 & -0.0025 & 0.000022 \end{bmatrix} \begin{bmatrix} 13.7 \\ 2.46 \\ 2.98 \\ 1116 \end{bmatrix}$$

$$\frac{1}{2} \hat{\mu}_1^T \hat{S}^{-1} \hat{\mu}_1 = 432$$

TABLE 4.15   Prediction of Alcohol Class Using the Three Membership Functions

Malic acid	Alkalinity of ash	Nonflavanoids	Proline	Alcohol	f_1	f_2	f_3	Prediction
14.23	2.43	3.06	1065	1	232	230	229	1
13.2	2.14	2.76	1050	1	193	192	190	1
13.16	2.67	3.24	1185	1	200	198	197	1
...	...	...	...	...	...	...	...	...
12.37	1.36	0.57	520	2	166	170	168	2
12.33	2.28	1.09	680	2	182	184	183	2
12.64	2.02	1.41	450	2	198	200	199	2
...	...	...	...	...	...	...	...	...
13.27	2.26	0.69	835	3	200	200	202	3
13.17	2.37	0.68	840	3	200	200	202	3
14.13	2.74	0.76	560	3	252	252	255	3
...	...	...	...	...	...	...	...	...

To calculate the final piece of the equation, $\log(p_1)$:

$$\log(p_1) = \log(0.33) = -1.10$$

The final score for f_1 is:

$$f_1 = 449 - 432 - 1.10 = 232$$

Similarly, scores for f_2 and f_3 are calculated, which are 230 and 229, respectively, and the class corresponding to the largest score is selected as the predictive value ("1").

Table 4.15 illustrates the calculation of the scores for a number of the observations. The highest scoring function is assigned as the prediction. The contingency table in Fig. 4.28 details the cross-validated (using a 5% cross-validation) predictive accuracy of this model.

Discriminant anaysis model summary	
Descriptors	Malic acid, Alkalinity of ash, Nonflavanoids, Proline
Response	Alcohol

Cross validated results	
Accuracy	0.966
Error	0.0337

Actual (alcohol)

		1	2	3	Totals
Predicted (alcohol)	1	56	1	0	57
	2	3	68	0	71
	3	0	2	48	50
	Totals	59	71	48	178

Figure 4.28   Summary of the cross-validated discriminant analysis model

In the same manner as described in Section 4.3.8, different independent variable combinations can be assessed and the most promising approach selected. In addition, interaction terms (such as the product of variables) or other transformed variables, such as quadratic terms, can be incorporated into the model to help in nonlinear situations or if other assumptions are violated. Alternatively, the use of the quadratic discriminant functions would be appropriate (Hastie, 2003).

4.5 LOGISTIC REGRESSION

4.5.1 Overview

Logistic regression is a popular method for building predictive models when the response variable is binary. Many data mining problems fall into this category. For example, a patient contracts or does not contract a disease or a cell phone subscriber does or does not switch services. Logistic regression models are built from one or more independent variables, that can be continuous, discrete, or a mixture of both. In addition to classifying observations into these categories, logistic regression will also calculate a probability that reflects the likelihood of a positive outcome. This is especially useful in prioritizing the results. As an example, a marketing company may use logistic regression to predict whether a customer will or will not buy a specific new product. While the model may predict more customers than the company has resources to pursue, the computed probability can be used to prioritize the most promising candidates.

Unlike discriminant analysis, logistic regression does not assume that the independent variables are normally distributed, or have similar variance in each group. There are, however, a number of limitations that apply to logistic regression including (1) a requirement for a large data set with sufficient examples of both categories, (2) that the independent variables are neither additive nor collinear, and (3) that outliers can be problematic.

4.5.2 Logistic Regression Formula

Logistic regression usually makes predictions for a response variable with two possible outcomes, such as whether a purchase does or does not take place. This response variable can be represented as 0 and 1, with 1 representing the class of interest. For example, 1 would represent the "buy" class and 0 would represent the "does not buy" class. A formula is generated to calculate a prediction from the independent variables. Instead of predicting the response variable, the formula estimates a probability that the response variable is 1, or $P(y = 1)$. A standard linear regression formula would compute values outside of the $0-1$ range and is not used for this and other reasons. An alternative function is used in this situation. This function ensures the prediction is in the $0-1$ range, by following a sigmoid curve. This curve for a single independent variable is shown in Fig. 4.29. A logistic response function has the following formula:

$$P(y = 1) = \frac{1}{1 + e^{-(\beta_0 + \beta_1 x_1 + \beta_2 x_2 + \cdots + \beta_k x_k)}}$$

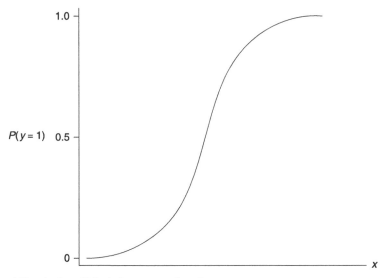

Figure 4.29   A sigmoid logistic response function

where β_0 is a constant, and $\beta_1 - \beta_k$ are coefficients to the k independent variables $(x_1 - x_k)$.

As an example, a logistic regression model is developed to predict whether a cereal would have a high nutritional rating using data from (http://lib.stat.cmu. edu/DASL/Datafiles/Cereals.html). It is based on the following logistic regression formula that uses measurements for variables *calories* (calories per serving), *protein* (grams of protein), and *carbo* (grams of complex carbohydrates):

$$P(y = 1) = \frac{1}{1 + e^{-(11.9 - 0.224 \times calories + 2.69 \times protein + 0.288 \times carbo)}}$$

For cereal A, which has *calories* = 90, *protein* = 3, *carbo* = 19, the predicted probability that this cereal would have a high nutritional rating is:

$$P(y = 1) = \frac{1}{1 + e^{-(11.9 - 0.224 \times 90 + 2.69 \times 3 + 0.288 \times 19)}}$$

$$P(y = 1) = 0.995$$

For cereal B with *calories* = 110, *protein* = 2, *carbo* = 12, the predicted probability is:

$$P(y = 1) = \frac{1}{1 + e^{-(11.9 - 0.224 \times 110 + 2.69 \times 2 + 0.288 \times 12)}}$$

$$P(y = 1) = 0.020$$

A cutoff is usually set such that probabilities above this value are assigned to class 1, and those below the cutoff are assigned to class 0. In this example, by setting a cutoff at 0.5, cereal A is assigned to the category high nutritional value, and cereal B is not.

"Odds" is a commonly used term, particularly in gambling. The odds are often referred to, for example, as "5 to 1" (such as to describe a bet), which translates into a 0.20 probability. The odds ratio considers $P(y = 1)$ vs $P(y = 0)$, using the following formula:

$$\text{odds} = \frac{P(y = 1)}{1 - P(y = 1)}$$

This allows us to rewrite the logistic regression formula as:

$$\text{odds} = e^{\beta_0 + \beta_1 x_1 + \cdots + \beta_i x_i}$$

This equation helps to interpret the coefficients of the equation. For an individual independent variable (x) with corresponding beta coefficient (β), holding all other variables constant, and increasing the value by 1 would result in the odds being increased by e^β. For example, in the cereal example, increasing the value of *carbo* by 1 would result in an increase in odds of being a high nutritional value cereal of by $e^{0.288}$ which is 1.33 or 33%.

It is also helpful to consider taking the natural log, which results in the following formula or *logit* function that will return a value between $-\infty$ and $+\infty$:

$$\log(\text{odds}) = \beta_0 + \beta_1 x_1 + \cdots + \beta_i x_i$$

4.5.3 Estimating Coefficients

The logistic regression coefficients are computed using a maximum likelihood procedure (Agresti, 2002), where the coefficients are continually refined until an optimal solution is found. The Newton–Raphson method is often used. Since the method is repeated multiple times, the estimated values for the coefficients $\hat{\beta}^{\text{new}}$ are updated using the previous estimates $\hat{\beta}^{\text{old}}$, based on the following formula:

$$\hat{\beta}^{\text{new}} = \hat{\beta}^{\text{old}} + (X^T W X)^{-1} X^T (y - p)$$

where X is the matrix describing the independent variables (with the first column assigned as 1 for calculation of the intercept), p is a vector of fitted probabilities, W is a weight matrix where the diagonal values represent $p(1 - p)$, and y is the response variable.

The method starts by assigning arbitrary values to the β coefficients. The β coefficients are repeatedly calculated using the formula above. Each iteration results in an improved coefficient estimate, and the process finishes when the beta coefficients are not changing significantly between iterations.

In the following example, a data set relating to diabetes is used to illustrate the process of calculating the coefficients (http://archive.ics.uci.edu/ml/datasets/Pima+Indians+Diabetes). Table 4.16 illustrates some of the data used to build the logistic regression model and Fig. 4.30 summarizes the data set. The *diabetes* variable is used as the response and the other five variables used as independent variables.

The following illustrates the process of calculating the β coefficients. In the first step, the β coefficients are initialized to an arbitrary value, in this case zero:

$$\hat{\beta}^{\,old} = \begin{bmatrix} 0.0 \\ 0.0 \\ 0.0 \\ 0.0 \\ 0.0 \\ 0.0 \end{bmatrix}$$

The X matrix is constructed from the independent variables, again, with the first column containing all 1s for calculation of the intercept value. The first three rows are shown:

$$X = \begin{bmatrix} 1 & 97 & 64 & 18.2 & 0.299 & 21 \\ 1 & 83 & 68 & 18.2 & 0.624 & 27 \\ 1 & 97 & 70 & 18.2 & 0.147 & 21 \\ \cdots & \cdots & \cdots & \cdots & \cdots & \cdots \end{bmatrix}$$

TABLE 4.16   Data Table Concerning Diabetes Information

Plasma glucose	Diastolic blood pressure	Body mass index	Diabetes pedigree function (DPF)	Age	Diabetes
97	64.0	18.2	0.299	21	0
83	68	18.2	0.624	27	0
97	70	18.2	0.147	21	0
104	76	18.4	0.582	27	0
80	55	19.1	0.258	21	0
99	80	19.3	0.284	30	0
103	80	19.4	0.491	22	0
92	62	19.5	0.482	25	0
100	74	19.5	0.149	28	0
95	66	19.6	0.334	25	0
129	90	19.6	0.582	60	0
162	76	49.6	0.364	26	1
122	90	49.7	0.325	31	1
152	88	50.0	0.337	36	1
165	90	52.3	0.427	23	0
115	98	52.9	0.209	28	1
162	76	53.2	0.759	25	1
88	30	55.0	0.496	26	1
123	100	57.3	0.88	22	0
180	78	59.4	2.42	25	1
129	110	67.1	0.319	26	1

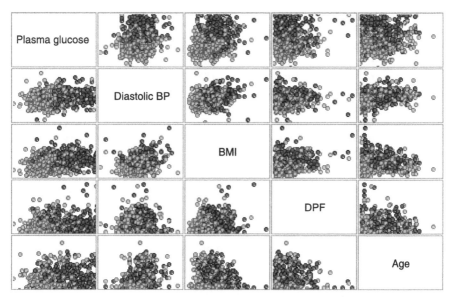

Figure 4.30   Scatterplot matrix with diabetes observations highlighted

The y matrix represents the response data, with the first three rows shown:

$$y = \begin{bmatrix} 0 \\ 0 \\ 0 \\ \ldots \end{bmatrix}$$

A p matrix corresponding to the calculated probability for each observation, using the current β coefficients, is calculated. The first three entries are shown here:

$$p = \begin{bmatrix} 0.5 \\ 0.5 \\ 0.5 \\ \ldots \end{bmatrix}$$

A weight matrix, W, where the number of columns and rows both equal the number of observations, is calculated. The diagonal represents $p(1 - p)$, and the first three rows and columns are shown here:

$$W = \begin{bmatrix} 0.25 & 0 & 0 & \ldots \\ 0 & 0.25 & 0 & \ldots \\ 0 & 0 & 0.25 & \ldots \\ \ldots & \ldots & \ldots & \ldots \end{bmatrix}$$

These matrices are used to generate an updated value for the beta coefficients:

$$\hat{\beta}^{\text{new}} = \hat{\beta}^{\text{old}} + (X^{\mathrm{T}} W X)^{-1} X^{\mathrm{T}} (y - p)$$

TABLE 4.17 Optimization of the β Coefficients for Logistic Regression

	β_0	β_1	β_2	β_3	β_4	β_5
Step 1	0.0	0.0	0.0	0.0	0.0	0.0
Step 2	−6.162	0.0247	−0.00404	0.0552	0.550	0.0231
Step 3	−8.433	0.0326	−0.00654	0.0814	0.836	0.0320
Step 4	−8.982	0.0344	−0.00733	0.0882	0.918	0.0343
Step 5	−9.009	0.0345	−0.00738	0.0886	0.923	0.0344
Step 6	−9.009	0.0345	−0.00738	0.0886	0.923	0.0344

$$\hat{\beta}^{\text{new}} = \begin{bmatrix} -6.162 \\ 0.0247 \\ -0.00404 \\ 0.0552 \\ 0.0552 \\ 0.0231 \end{bmatrix}$$

In Table 4.17, the first two rows illustrate the β coefficients calculated up to this point. This process is repeated until the coefficients values converge, that is, when the values do not change significantly between steps.

4.5.4 Assessing and Optimizing Results

Once a logistic regression formula has been generated, it can be used for prediction. In Table 4.18, two new columns are added. The logistic regression formula will calculate a probability, and from this value a classification can be assigned. A cutoff value, such as 0.5, can be used with a probability higher than 0.5 assigned to the 1 class and a probability less than or equal to 0.5 assigned to the 0 category. These classifications can be used to assess the model, using a contingency table, such as the one in Fig. 4.31.

The contingency table allows for the calculation of overall accuracy, error rate, specificity, sensitivity, and so on. In addition, since the model also generates a probability, a lift chart and an ROC chart can also be generated. Any models generated can be optimized by varying the independent variables to generate the simplest, most predictive model, as discussed in Section 4.3.8. The cutoff value can also be adjusted to enhance the quality of the model. References to additional methods for assessing the logistic regression model, such as the Wald test, the likelihood ratio test, and Hosmer and Lemeshow χ^2 test of goodness of fit are provided. Like linear regression and discriminant analysis, interaction terms (such as the product of two variables) or higher-order terms (such as a variable squared) can be computed and used with the model to enhance prediction. Chapter 5 illustrates the application of logistic regression to a number of case studies.

TABLE 4.18   Prediction of Diabetes as Well as the Probability of Diabetes is 1

Plasma glucose	Diastolic blood pressure	Body mass index	Diabetes pedigree function (DPF)	Age	Diabetes	Predicted	Probability
97	64.0	18.2	0.299	21	0	0	0.0287
83	68	18.2	0.624	27	0	0	0.0285
97	70	18.2	0.147	21	0	0	0.0240
104	76	18.4	0.582	27	0	0	0.0530
80	55	19.1	0.258	21	0	0	0.0180
99	80	19.3	0.284	30	0	0	0.0425
103	80	19.4	0.491	22	0	0	0.0371
92	62	19.5	0.482	25	0	0	0.0365
100	74	19.5	0.149	28	0	0	0.293
95	66	19.6	0.334	25	0	0	0.0263
129	90	19.6	0.582	60	0	0	0.0416
162	76	49.6	0.364	26	1	1	0.839
122	90	49.7	0.325	31	1	1	0.577
152	88	50.0	0.337	36	1	1	0.828
165	90	52.3	0.427	23	0	1	0.863
115	98	52.9	0.209	28	1	1	0.520
162	76	53.2	0.759	25	1	1	0.909
88	30	55.0	0.496	26	1	1	0.508
123	100	57.3	0.88	22	0	1	0.759
180	78	59.4	2.42	25	1	1	0.993
129	110	67.1	0.319	26	1	1	0.854

		Actual (diabetes)		
		1	0	Totals
Predicted (diabetes)	1	140	55	195
	0	109	420	529
	Totals	249	475	724

Figure 4.31   Contingency table summarizing the number of correct and incorrect predictions

4.6 NAIVE BAYES CLASSIFIERS

4.6.1 Overview

The *naive Bayes* (also referred to as idiot's Bayes or simple bayesian classifier) is a classification modeling method. It makes use of the *Bayes theorem* to compute probabilities of class membership, given specific *evidence*. In this scenario, the evidence refers to particular observations in the training set that either support or do not support a particular prediction.

Naive Bayes models have the following restrictions:

- *Only categorical variables*: This method is usually applied in situations in which the independent variables and the response variable are categorical.

- *Large data sets*: This method is versatile, but it is particularly effective in building models from large data sets.

This method provides a simple and efficient approach to building classification prediction models. It also can compute probabilities associated with class membership, which can be used to rank the results.

4.6.2 Bayes Theorem and the Independence Assumption

At the heart of this approach is the Bayes theorem:

$$P(H|E) = \frac{P(E|H)P(H)}{P(E)}$$

This theorem calculates the probability of a *hypothesis* (H) given some evidence (E), or *posterior probability*. For example, it can calculate the probability that someone would develop diabetes given evidence of a family history of diabetes. The hypothesis corresponds to the response variable in the other methods. The theorem makes use of this posterior probability of the evidence given the hypothesis, or $P(E|H)$. Using the same example, the probability of someone having a family history of diabetes can also be calculated given the evidence that the person has diabetes and would be an example of $P(E|H)$. The formula also makes use of two prior probabilities, the probability of the hypothesis $P(H)$, and the probability of the evidence $P(E)$. These probabilities are not predicated on the presence of any evidence. In this example, the probability of having diabetes would be $P(H)$, and the probability of having a family history of diabetes would be $P(E)$.

4.6.3 Independence Assumption

In situations in which there is only a single independent variable, the formula is straightforward to apply; however, models with a single independent variable would be limited in their usefulness. Unfortunately, the strict use of the Bayes theorem for multiple independent variables each having multiple possible values becomes challenging in practical situations. Using this formula directly would result in a large number of computations. Also, the training data would have to cover all of these situations, which also makes its application impractical. The naive Bayes approach uses a simplification which results in a computationally feasible series of calculations. The method assumes that the independent variables are independent despite the fact that this is rarely the case. Even with this overly optimistic assumption, the method is useful as a classification modeling method in many situations.

TABLE 4.19   Diabetes Data Set to Illustrate the Naive Bayes Classification

Blood pressure	Weight	Family history	Age	Diabetes
Average	Above average	Yes	50+	1
Low	Average	Yes	0–50	0
High	Above average	No	50+	1
Average	Above average	Yes	50+	1
High	Above average	Yes	50+	0
Average	Above average	Yes	0–50	1
Low	Below average	Yes	0–50	0
High	Above average	No	0–50	0
Low	Below average	No	0–50	0
Average	Above average	Yes	0–50	0
High	Average	No	50+	0
Average	Average	Yes	50+	1
High	Above average	No	50+	1
Average	Average	No	0–50	0
Low	Average	No	50+	0
Average	Above average	Yes	0–50	1
High	Average	Yes	50+	1
Average	Above average	No	0–50	0
High	Above average	No	50+	1
High	Average	No	0–50	0

4.6.4 Classification Process

To illustrate the naive Bayes classification process, the training set in Table 4.19 will be used to classify the following observation (X):

$$X : BP = high;\ weight = above;\ FH = yes;\ age = 50+$$

In this example, the observation (X) is an individual whose blood pressure is high $(BP = high)$, whose weight is above normal $(weight = above)$, who has a family history of diabetes $(FH = yes)$, and whose age is above 50 $(Age = 50+)$. Using the training data in Table 4.19, the objective is to classify this individual as prone to developing or not prone to developing diabetes given the factors described. In this example, calculating $P(diabetes = 1|X)$ and the $P(diabetes = 0|X)$ is the next step. The individual will be assigned to the class, either has $(diabetes = 1)$ or has not $(diabetes = 0)$, based on the highest probability value.

$$P(diabetes = 1|X) = \frac{P(X|diabetes = 1)P(diabetes = 1)}{P(X)}$$

$$P(diabetes = 0|X) = \frac{P(X|diabetes = 0)P(diabetes = 0)}{P(X)}$$

Since $P(X)$ is the same in both equations, only $P(X|diabetes = 1)P(diabetes = 1)$ and $P(X|diabetes = 0)P(diabetes = 0)$ are needed.

To calculate $P(diabetes = 1)$, the number of observations is counted in Table 4.19 with $diabetes = 1$, which is 9, divided by the total number of observations, which is 20:

$$P(diabetes = 1) = 9/20 = 0.45$$

Similarly, to calculate $P(diabetes = 0)$, the number of observations in Table 4.19 is counted where $diabetes = 0$, which is 11, divided by the total number of observations, which is 20:

$$P(diabetes = 0) = 11/20 = 0.55$$

Since this approach assumes that the independent variables are independent, the calculation of the $P(X|diabetes = 1)$ is the product of the conditional probability for each of the values of X:

$$P(X|diabetes = 1) = P(BP = high|diabetes = 1)$$
$$\times P(weight = above|diabetes = 1)$$
$$\times P(FH = yes|diabetes = 1)$$
$$\times P(age = 50+|diabetes = 1)$$

The individual probabilities are again derived from counts of Table 4.19. For example, $P(BP = high|diabetes = 1)$ counts all observations with $BP = high$ and $diabetes = 1(4)$, divided by the number of observations where $diabetes = 1(9)$:

$$P(BP = high|diabetes = 1) = 4/9 = 0.44$$
$$P(weight = above|diabetes = 1) = 7/9 = 0.78$$
$$P(FH = yes|diabetes = 1) = 6/9 = 0.67$$
$$P(age = 50+|diabetes = 1) = 7/9 = 0.78$$

Using these probabilities, the probability of X given $diabetes = 1$ is calculated:

$$P(X|diabetes = 1) = P(BP = high|diabetes = 1)$$
$$\times P(weight = above|diabetes = 1)$$
$$\times P(FH = yes|diabetes = 1)$$
$$\times P(age = 50+|diabetes = 1)$$
$$P(X|diabetes = 1) = 0.44 \times 0.78 \times 0.67 \times 0.78$$
$$P(X|diabetes = 1) = 0.179$$

Using the values for $P(X|diabetes = 1)$ and $P(diabetes = 1)$, the product $P(X|diabetes = 1)P(diabetes = 1)$ can be calculated:

$$P(X|diabetes = 1)P(diabetes = 1) = 0.179 \times 0.45$$
$$P(X|diabetes = 1)P(diabetes = 1) = 0.081$$

Similarly, the value for $P(X|diabetes = 0)P(diabetes = 0)$ can be calculated:

$$P(X|diabetes = 0) = P(BP = high|diabetes = 0)$$
$$\times P(weight = above|diabetes = 0)$$
$$\times P(FH = yes|diabetes = 0)$$
$$\times P(age = 50 + |diabetes = 0)$$

Using the following probabilities, based on counts from Table 4.19:

$$P(BP = high|diabetes = 0) = 4/11 = 0.36$$
$$P(weight = above|diabetes = 0) = 4/11 = 0.36$$
$$P(FH = yes|diabetes = 0) = 4/11 = 0.36$$
$$P(age = 50+|diabetes = 0) = 3/11 = 0.27$$

The $P(X|diabetes = 0)$ can now be calculated:

$$P(X|diabetes = 0) = 0.36 \times 0.36 \times 0.36 \times 0.27$$
$$P(X|diabetes = 0) = 0.0126$$

The final assessment of $P(X|diabetes = 0)P(diabetes = 0)$ is computed:

$$P(X|diabetes = 0)P(diabetes = 0) = 0.0126 \times 0.55 = 0.0069$$

Since $P(X|diabetes = 1)P(diabetes = 1)$ is greater than $P(X|diabetes = 0)$ $P(diabetes = 0)$, the observations X are assigned to class $diabetes = 1$. A final probability that $diabetes = 1$, given the evidence (X), can be computed as follows:

$$P(diabetes = 1|X) = 0.081/(0.081 + 0.0069) = 0.922$$

The naive Bayes is a simple classification approach that works surprisingly well, particularly with large data sets as well as with larger numbers of independent variables. The calculation of a probability is helpful in prioritizing the results. As with other methods described in the chapter, the predictive accuracy of any naive Bayes model can be assessed using the methods outlined in Section 4.1. Building models with different sets of independent variables can also help.

4.7   SUMMARY

The preceding chapter has discussed two basic types of models:

- *Classification*: model where the response is a categorical variable;
- *Regression*: model where the response is a continuous variable.

TABLE 4.20   Summary of Methods to Assess Regression Models

Mean square error:

$$\frac{\sum_{i=1}^{n} (\hat{y}_i - y_i)^2}{n}$$

Mean absolute error:

$$\frac{\sum_{i=1}^{n} |\hat{y}_i - y_i|}{n}$$

Relative square error:

$$\frac{\sum_{i=1}^{n} (\hat{y}_i - y_i)^2}{\sum_{i=1}^{n} (y_i - \bar{y})^2}$$

Relative absolute error:

$$\frac{\sum_{i=1}^{n} |\hat{y}_i - y_i|}{\sum_{i=1}^{n} |y_i - \bar{y}|}$$

Correlation coefficient:

$$\frac{\sum_{i=1}^{n} (y_i - \bar{y})(\hat{y}_i - \bar{\hat{y}})}{(n-1)s_y s_{\hat{y}}}$$

TABLE 4.21   Summary of Methods to Assess Binary Classification Models

Accuracy:	Error rate:	Sensitivity:	Specificity:
$\dfrac{TP + TN}{TP + FP + FN + TN}$	$1 - \dfrac{TP + TN}{TP + FP + FN + TN}$	$\dfrac{TP}{TP + FN}$	$\dfrac{TN}{TN + FP}$
False positive rate:	Positive predictive value:	Negative predictive value:	False discovery rate:
$\dfrac{FP}{FP + TN}$	$\dfrac{TP}{TP + FP}$	$\dfrac{TN}{TN + FN}$	$\dfrac{FP}{FP + TP}$

TABLE 4.22   Summary of Predictive Modeling Methods Discussed in this Chapter

Method	Model type	Independent variables	Comments
Multiple linear regression	Regression	Any numeric	Assumes a linear relationship Easy to explain Quick to build
Discriminant analysis	Classification	Any numeric	Assumes the existence of mutually exclusive groups with common variances
Logistic regression	Classification	Any numeric	Will calculate a probability Easy to explain Limit on number of observations to build models from
Naive Bayes	Classification	Only categorical	Requires a lot of data

Multiple methods have been discussed for assessing regression and classification models, and are summarized in Tables 4.20 and 4.21. Principal component analysis has been discussed as a method for understanding and restricting the number of variables in a data set. Table 4.22 summarizes the methods discussed in this chapter.

4.8 FURTHER READING

For further general discussion on the different methods for assessing models see Han and Kamber (2005) and Witten and Frank (2005). Here additional validation methods, such as bootstrapping, are discussed, along with methods for combining models using techniques such as bagging and boosting. Joliffe (2002), Strang (2006), Jackson (1991), Jobson (1992), and Johnson and Wishern (1998) provide additional detail concerning principal component analysis. Additional information on multiple linear regression can be found in Allison (1998), Draper and Smith (1998), Fox (1997), and Rencher (2002). Discriminant analysis is also covered in more depth in Hastie et al. (2003), McLachlan (2004), Huberty (1994), Lachenbruch (1975), and Rencher (2002). Agresti (2002), Balakrishnan (1992), and Hosmer and Lemeshow (2000) cover logistic regression, and Han and Kamber (2005) as well as Hand and Yu (2001) discuss the use of the naive Bayes approach. Other commonly used methods for building prediction models include neural networks (Hassoun, 1995; Haykin, 1998; and Myatt, 2007), classification and regression trees, rule-based classifiers, support vector machines, and k-nearest neighbors, and these are covered in a variety of books, including Han and Kamber (2005), Witten and Frank (2005), Hastie et al. (2003), and Shumueli et al. (2007).

CHAPTER **5**

APPLICATIONS

5.1 OVERVIEW

Data mining is being increasingly used to solve a variety of business and scientific problems. Ideally, its use is tied to achieving specific strategic objectives, such as developing a more personalized relationship with existing customers. It should also be well integrated within new or existing business processes, which are generally accepted. Careful consideration should be given to what and how specific information is collected, potentially utilizing a data warehouse to store this decision-support repository. Which data mining methods are used is just one issue within a host of concerns that need to be addressed. It is also important to address how the results are communicated and used, such as through an embedded system, the generation of reports, or through online tools. Continually monitoring the business impact of the data mining exercise is critical to ensuring the success of any project. Those organizations that are using data mining within this context are seeing a significant return on their data mining investment.

A review of how data mining is being used across a broad range of industries indicates a number of common problems are being addressed. Data mining applications are now commonly used to accomplish the following:

- *Enhanced operational efficiency*: Data mining can facilitate the efficient allocation of resources. For example, government departments are planning future fire stations and other resources based on an analysis of historical fire incidents.
- *Improved marketing campaigns*: Data mining enables organizations to build relationships directed towards individual customer groups, allowing them to acquire and retain customers more easily, along with planning the launch and promotion of new or existing products.
- *Management of risk*: Data mining can be used to make predictions about future events, facilitating more informed decisions. For example, insurance companies are using data mining solutions to more effectively underwrite policies.
- *Detection of problems*: Early detection of errors and sources of problems can be found through the use of data mining. Telecommunications companies, for example, are using data mining to understand and circumvent problems across their networks.

Making Sense of Data II. By Glenn J. Myatt and Wayne P. Johnson
Copyright © 2009 John Wiley & Sons, Inc.

165

- *Identification of fraud*: Data mining can be used to identify factors that are related to fraudulent activity and used to improve detection. As an example, tax collection agencies are using data mining to understand patterns associated with individuals who do not pay.

- *Support for research and development*: Data mining can be used to sift through large databases to help develop new products or advance the science. For example, pharmaceutical companies are identifying patterns and trends in large quantities of research data, using data mining techniques, to find new drugs.

Data mining is also being used across different functional areas within organizations. For example, human resource departments are using data mining to model the employee experience life cycle, and these models are used to enhance employee retention. Within planning departments, data mining is being utilized to identify new business locations, as well as to support merger and acquisition activities. Customer support is using data mining to initially classify incoming service requests from the web for triage purposes.

Data mining is being increasingly used by data mining experts as well as subject-matter specialists. Methods such as visual data mining are used to see trends in the data without the need for data mining training. For example, production workers are able to visually analyze data generated from a manufacturing production and quickly take care of any issues identified.

Data mining techniques are also being embedded into other applications. For instance, SPAM filtering makes extensive use of data mining approaches, such as decision trees, to partition incoming emails into folders. Models are continuously built from historical data where emails have been classified as SPAM and not SPAM. These models are then incorporated into the email system, where incoming emails are automatically predicted, and those determined to be SPAM are set aside.

The following, while not comprehensive, illustrates the breadth of applications to which data mining has been applied. The use of data mining for sales and marketing activities is common to many industries, and its use is generally described. In addition, data mining is used to solve many problems in a diverse range of industries. This chapter outlines the use of data mining in the following industries: finance, insurance, retail, telecommunications, manufacturing, entertainment, government, pharmaceuticals, and healthcare. In addition, two case studies are described that outline data mining projects in the area of microRNA analysis and scoring credit loans applications. Both case studies use the Traceis software that is available from http://www.makingsenseofdata. com/ and described in Appendix B. This chapter also discusses the use of data mining in situations where the data is not in a tabular format, and such data needs to be preprocessed to make the information amenable to data mining methods. The use of preprocessing data to facilitate data mining is described in the context of two examples: data mining chemical information and data mining text.

5.2 SALES AND MARKETING

One of the widespread uses of data mining is in the area of sales and marketing. Organizations are data mining information collected from customer purchases, supplemented with additional information such as usage, lifestyle, demographic, and geographical data. This use has led to improvements in how new customers are identified or prioritized, how current customers are retained, and how business may be expanded with existing clients. Data mining solutions can also help to model how customers purchase products and services over time and how to effectively customize online retailing. The following section summarizes some of the ways data mining has been used to improve an organization's sales operations:

- *Acquiring new customers*: Companies spend a great deal of time and money acquiring new customers. Since these costs are high, data mining technology is used to help ensure a return on a company's investment in customer acquisition. Predictive analytics help focus marketing outreach campaigns, such as the promotion of a new product or a special offer towards those prospects with the highest chance of becoming customers. Predictive models are usually built from historical sales data and other information collected about the target customer, such as age or location. The models attempt to understand the relationships between specific customer attributes and the likelihood that those types of people will become customers. These models are built to rank customers from lists of prospective customers. Other factors relating to customer acquisition may also be built into the model. For example, a customer may fit the profile of a customer likely to purchase but may also be a high credit risk or expensive to support in a long-term business relationship. In addition, prospects can be targeted with specific products they are predicted to need, which in turn further increases the chance of acquiring those customers.

- *Retaining customers*: Companies are at risk of losing a significant amount of money if a profitable customer decides to switch to another institution. This problem is often called *attrition* or *churn*, and data mining approaches are being used to identify and prioritize "at risk" customers. Prediction models are often built to predict the probability of losing specific customers. These models are built from data sets containing information for given time periods—such as a month—on transaction volumes, products and services utilized, and personal demographics. A binary response variable is often used to indicate whether a customer terminates an account within one of the time periods. One problem when building these types of models is the imbalance between the often low numbers of examples where customers changed services compared to the majority of cases in which the customers did not. This imbalance can present difficulties in building models, and hence a sampling method to balance these two cases is commonly used. Modeling techniques, such as logistic regression, are used to build such models. Models are also built to minimize false positives because it is ultimately more costly to lose a customer as opposed to providing a financial incentive so a customer does not considering switching services. These models can then be used to prioritize customers based on those with the highest probability of moving to a different company, and the company can then target those customers with marketing activities, such as promotional offers. Other factors, such as the profitability of the customer, can also be built into these or other models to help focus customer retention resources. For example, the telecommunications industry attempts to keep churn levels to a minimum using data mining. The problem in this industry is particularly challenging as a result of the difficulty in customer acquisitions because the market is somewhat saturated. This means that the cost of retaining a customer is significantly less than the cost of acquiring a new customer. Data concerning call details, along with billings, subscriptions, and demographic information, collected over a given time period is used to build the models. Understanding this behavior allows the telecommunications company to focus marketing campaigns, offering new services that specifically focus on customer retention.

- *Increasing business from existing customers*: The ability to sell new products and services to existing customers is a significant focus in many companies' sales and marketing divisions. It is usually more profitable to sell additional products and services to existing customers than to acquire new customers. Information on customer transactions is supplemented with data about customer demographics and lifestyle, as well as geographical information. This information may be obtained from a loyalty program application or from a third-party supplier. Aggregated information from historical data, such as the propensity to respond to marketing campaigns, is often added. Data transformations are often applied to ensure the relevant business questions are answered, for example, using product hierarchies that generalize individual purchases into general categories. This rich collection of information is often data mined to understand customer segments associated with certain buying habits and customer preferences. These groups can also help in understanding those segments of the market that are most profitable. This segmentation enables highly focused marketing campaigns to promote products and product combinations to specific consumer groups in specific locations. These targeted promotions result in significantly higher response rates than tradition blanket marketing campaigns, at a reduced cost. Organizations often tightly integrate these analysis capabilities within their customer relationship management (CRM) systems and the models are continually checked to ensure they are up-to-date. These models are often tested using smaller scale trial runs, and the results of the trial runs can be used to determine whether the models will be used in a more expansive marketing campaign.

- *Understanding sales channels*: Predictive models are also built to identify the sales channel through which any offer should be made, such as through the internet, the mail, through a branch office, or even through specialized consultants. Models built to predict the appropriate sales channel are based on the profile of the sales lead.

- *Modeling the customer experience*: Linking transaction data to the identity of a customer is often achieved through a loyalty program, such as a loyalty card, that the consumer uses on each visit to a retail or grocery store. This information allows the organization to understand the buying patterns of a customer over time, or how that individual shops, that is, whether the person shops online or through other retail channels. Over time, companies generate profiles of how a customer shops which allow them to model a customer's purchasing cycle over time, and this makes for more effective timing of promotional activities. For example, publishing companies are using these models to understand how customers move from one magazine title to another over time, helping the publishers keep customers within their product suites.

- *Personalizing online retailing*: Data mining has been used extensively in the area of online retailing, capitalizing on the wealth of digital information that is continually collected. In addition to sales transaction data, information is collected from customer ratings and reviews, the web pages viewed, and information on current product inventory levels. Clustering is often used to generate segments of users, and these groups are annotated with information on product popularity and overall profitability. Association rules are also employed to identify product combinations that are often purchased together. These results are used to personalize web pages based on the profile of the customer built over time. They are also matched to similar customers' preferences. This personalization includes products the consumer is anticipated to be interested in and educational material, such as white papers. It also makes use of models that prioritize profitable customers. This personalization often takes into account inventory levels to maximize conversion rates and profits. Online retailing offers a fertile ground for developing new and novel methods for cross- and up-selling. Ensuring that the data mining results in profitable updates to the online experience is easily tested in the online environment. Subscribers are often divided into groups that make use of a new service and a control group that does not have access to the service. The success of any new service is based on differences recorded for the two groups, such as higher sales or profits.

5.3 INDUSTRY-SPECIFIC DATA MINING

5.3.1 Finance

The financial industry encompasses a wide range of institutions including banks, asset management companies, and accounting services. Data mining is used throughout the financial industry, driven by the intense competitive pressure between financial companies, as well as the availability of large volumes of data collected concerning customers, transactions, and the financial markets. Data mining approaches are being used to improve sales through more effective acquisition and retention of customers. These methods are also used to optimize cross-selling and up-selling marketing campaigns. In addition, the ability to manage risk is pervasive throughout the industry. For example, risk assessment is necessary to make loans or manage stock portfolios. Data mining is often used to lower risk in a wide variety of financial situations. Finally, it is also used to help in identifying criminal activities, such as credit card fraud.

- *Sales and marketing*: The financial industry's customers are often long-standing, and profits are made from this relationship over time. The institution is at risk of losing a significant amount of money if a profitable customer decides to switch to another institution. Many financial institutions offer a range of products and services, from loans and credit cards to wealth management and insurance services. The profitability of an individual customer is driven, to a large part, by the institution's ability to provide that customer with additional products and services over time. Financial companies spend a great deal of resources acquiring, retaining, and growing new customers, using methods described in Section 5.2 on sales and marketing.

- *Managing credit risk*: Financial institutions continually receive applications for loans, credit cards, and so on. They need to make a timely risk assessment on whether to provide these services to an applicant and under what terms. Standard credit scores are available based on credit histories; however, additional models are often generated that provide a higher degree of accuracy. These new models are often tailored to an individual product. In addition, customers are often segmented into groups, using clustering methods and specific credit assessment models generated for these typical and atypical groups. Models of credit risk are often built using methods such as logistic regression, decision trees, or neural networks. These types of models can be used during both the application process and the collection process. These models are often built under a general assumption, such as the current condition of the economy, and may need to be rebuilt should any of these assumptions change. Section 5.5 provides a case study to illustrate this scenario.

- *Identifying suspicious activity*: Detecting credit card fraud is a major challenge and various analytical approaches have been used to identify and circumvent this suspicious activity. Predictive models are built from historical transaction data including both fraudulent and nonfraudulent transactions, selecting observations with a balance of the two cases. Models are often built that maximize accuracy, and a level of false positives is tolerated. This is because, overall, the institutions wish to be able to identify as much fraudulent activity as possible. Models for other suspicious activities are also built and are used to detect criminal activities such as money laundering.

- *Mining trading data*: Data mining approaches, such as predictive analytics, are often applied to financial market data. Generating models and understanding trends in the financial markets are important tools to optimize investment profits. These data mining approaches are used to predict when to buy and sell stocks, and other investments. Multidimensional time series data sets are often used to build these prediction models. The data contains lots of noise and is invariably linked to major news events, which are sometimes even built into the model. Many different data mining approaches have been applied to this problem. Neural networks are extensively used since they handle the inherent noise in the data well. Although predictions over the long term are extremely difficult, predictions for short-term events are common. However, the usefulness of these models is often short-lived, and they need to be retrained constantly to provide effective predictions. In addition to predicting individual stock prices, data mining methods are used to assess portfolios of products, selecting optimal combinations.

5.3.2 Insurance

Insurance carriers, agencies, and brokerages carry a wide range of products, including business owner's insurance, health insurance, automobile insurance, and general liability insurance. The industry is highly competitive and companies often compete on price and service. Insurance providers need to make rapid assessments of risk when they provide insurance quotes. Insurance companies are increasingly providing quotes to potential customers in real time through internet sites. These underwriting activities are based on the risks associated with likely future events. To support these decisions, financial institutions have collected massive amounts of data over the years and use data mining to facilitate sales and marketing activities, underwrite policies, and accelerate legitimate claims processing, at the same time as avoiding fraudulent claims.

- *Growing business opportunities*: Insurance companies use a variety of channels to identify new customers, and predictive analytics have been used to improve customer acquisition and retention using methods detailed in Section 5.2 on sales and marketing.

- *Identifying health risks*: Particularly in the area of health insurance, data mining helps identify customers at risk of a future health problem based on information collected concerning the customer over time. These customers are approached with disease management options with the hope that these intervention programs avoid the onset of the predicted health issue or any complications.

- *Identifying new products*: Insurance companies can grow business and improve sales and profits by supplementing existing products with new products or by expanding into new geographical territories. Just as using segmentation and predictive models can help sell existing products, these techniques can also help identify new products and quickly point out the associated risks. This allows an insurance company to rapidly move into new insurance markets, which, in turn, creates a competitive advantage. In addition, decisions concerning the future allocation of resources, for example, which products to sell in what new areas, can be supported with data mining approaches. Internal underwriting data, along with external data, such as census data, industry data, economic, and other information concerning the population, commerce, and the competition in specific geographical areas, is collected. This data is then used to determine future strategies and office locations. Models developed can help to clarify the amount of money to invest along with specific product categories to focus on.

- *Improving underwriting*: An insurance underwriter must assess the risk associated with a specific policy and generate a premium based on this risk. Setting the price too high may cost the insurance company business; setting the price too low may undermine the insurance company's profits. Data mining approaches have been successfully applied to the process of underwriting, enabling more consistent pricing with higher profits. This is achieved by avoiding historically subjective assessment and taking into account a larger number of factors. In addition, customer populations are often segmented using demographic and geographical information, along with other data from internal and external sources. Predictive models are then built within these individual segments to estimate the relative risk associated with an individual policy. Understanding the factors leading to these segmented groups allows the insurer to maximize profits in each of these areas. Insurers offering online services also benefit from utilizing underwriting predictive models, thus allowing a prospective online customer to determine on-the-spot whether to purchase insurance based on the price, generated from the predictive models.

- *Identifying fraudulent activities*: Insurance fraud results from a number of situations, including inflated claims, misrepresented facts, and claims for nonexistent injuries or damages. Effective fraud detection would lead to significant financial benefits to the industry. Successful fraud detection would avoid the costs associated with processing fraudulent claims and allow legitimate claims to be processed faster. Today, only a small percentage of fraudulent cases are identified, and hence there is a need for more effective detection of these cases. Historically, fraud detection is performed while processing claims using specific rules that suggest a claim might be fraudulent. These rules were manually developed by specialists who attempt to discern factors that differentiate fraudulent claims from legitimate claims. The rules identified were only able to identify a small handful of cases. Data mining is now being used to improve the quality of these rules. Rules developed in conjunction with data mining techniques can also be constantly updated to reflect changes in illegitimate activity over time. The improved rules help to identify more fraudulent cases, in addition to accelerating the processing of those cases predicted to have a low probability of fraud. Any modeling effort, however, faces a number of challenges. Since the investigation of fraudulent claims is expensive and time-consuming, any prediction must attempt to maximize detection with minimal false positives. The availability of data also presents a challenge, since the number of fraudulent cases currently detected and prosecuted is low. Either these small data sets are used or data sets containing significant amounts of noise need to be used. This is because the larger number of "positive" cases has not been fully investigated, and hence will include cases that are not fraudulent.

5.3.3 Retail

Retail companies are continually striving to understand their target market in detail. This, in turn, enables them to offer the right mix of products to specific consumers. It also enables them to time correctly the availability of products and price them appropriately. Retail companies use a wide range of methods to gather information on their customers, and they are increasingly using data mining approaches to make decisions based on trends identified in historical databases. They have collected an enormous amount of information on point-of-sale, demographic, geographical, and other information. This information is mined to optimize profits throughout all retail channels, such as online or in stores. Information derived can be used in many situations such as up-selling and cross-selling promotional activities, identifying new store locations, and optimizing the entire supply chain.

- *Mining retail transactions data*: Every time a customer visits a retail store and buys something, information is collected concerning the visit and the purchased products. This data is often supplemented with information concerning local competitors' pricing, and all this data can be mined in a number of ways. Techniques, such as association rules, are used to identify combinations of products that are purchased together. This, then, helps with product placements, by grouping products commonly sold together. Promotional activities, for example using coupons to promote combinations of items, are additionally used to encourage cross-selling. This transaction information, especially when combined with competitive information, is often used to optimize prices. Using price elasticity models, retailers can dynamically adjust prices, including adjusting mark-downs in order to maximize profits. The transaction data is also useful in understanding purchasing trends over time in both the local and wider markets. This analysis will, in turn, help the retailer gain a better understanding of any seasonal changes.

- *Improving operational efficiency*: Projecting demand allows retail companies to manage inventory levels, avoiding costly overstocking while at the same time ensuring that products are available for their customers. In addition to predicting demand, other logistic information is also incorporated into the analysis in order to optimize the entire supply chain. This includes information on the transportation and distribution network, as well as information concerning the manufacturers. Models are created to optimize the entire process, resulting in lower costs. Models that predict future demand for products can also be useful input when negotiating with suppliers.

5.3.4   Telecommunications

The telecommunications industry refers to companies that deliver information such as voice, data, graphics, and videos through traditional wire lines, and increasingly through wireless services. Telecommunications companies are additionally offering cable and satellite program distribution as well as high-speed internet services. The combination of an increasing number of companies entering the market, coupled with excess capacity, has led to intense competition and consolidation within the industry. The industry generates massive amounts of data concerning its operations. For instance, individual calling information and other customer information is collected, and is being mined to provide a competitive advantage. These advantages include better management of costs as well as more effective customer retention, cross- and up-selling, and customer acquisition, as described in Section 5.2 on sales and marketing.

- *Detecting problems*: The equipment making up the networks used by telecommunications companies is complex and interrelated. The equipment automatically generates information concerning its status, that is, whether it is operating normally or there is a problem. The data collected concerning the functioning of equipment across the entire telecommunications network is collected and data mined to determine recurrent patterns of faults, and used to enhance the reliability of the network. This data is often prepared such that each observation is based on a specific fixed-length period of time, and then used within standard classification data mining approaches.

- *Assessing credit risk*: Predictive analytics are being increasingly used to make more accurate assessments of credit risk allowing the company to focus late-payment inquiries on those not likely to pay, while not targeting unnecessarily those who have not paid for other reasons, such as an unresolved payment issues.

- *Detecting fraud*: Data concerning calls (both fraudulent and legitimate), along with general customer data, are also used to identify fraudulent patterns. These patterns are used to screen for illegitimate use, and those calls deemed highly suspicious are alerted to an investigator. A number of issues must be addressed when preparing this type of data for use within a predictive model. Information on individual calls is usually summarized into observations that describe information such as the average duration of the call or average number of calls per day. Also, data on fraudulent activity is not evenly distributed between fraudulent and nonfraudulent examples, and often a stratification strategy is adopted to even out the two cases.

5.3.5 Manufacturing

The manufacturing industry is comprised of companies that produce new products, utilizing raw materials, substances or components. The manufacturing industry includes companies that produce a broad range of goods including aerospace products, semiconductors, apparel, motor vehicles, chemicals, and electronic products, just to name a few. As a consequence of industry consolidations, a relatively small number of companies employ a large number of employees, and these companies invariably compete globally. In addition, these companies have invested heavily in complex automated manufacturing system as well as information technology in order to stay competitive. Data is collected and analyzed concerning the operations of the manufacturing processes as well as sales and supply-chain data. Enterprise resource planning (ERP) systems are routinely used in this context. Data mining is being increasingly used to enhance the development of new products, to ensure the manufacturing process operates efficiently and with a high level of quality, as well as to meet the demands of customers in the marketplace.

- *Enhancing product design*: Data mining has been applied to the product design phase, where historical data along with data from other sources is used to optimize design specifications and control product component costs. This application typically involves multidimensional optimizations to identify the optimal configuration and list of suppliers.

- *Enhancing production quality*: During the manufacturing process, data is collected from control systems and can be mined to ensure consistency and detect problem sources. Data mining approaches enable the monitoring of the many systems that are common within typical manufacturing environments. These methods may also be embedded into the manufacturing system and used to optimize entire production processes, as well as detect potential errors or problems automatically. The information can also be used by production workers, often using data visualizations, to make production assessments.

- *Allocating factory resource*: Manufacturing data is often used to maximize layouts as well as to optimize the use of other resources such as energy, human resources, and suppliers. Models are also built to identify optimal machine settings used within the production process, as well as models that enable the most efficient scheduling of production events. Data mining can also be used with this enterprise information to avoid health and safety issues, and minimize any environmental impact resulting from the production.

- *Sales and maintenance*: Manufacturing companies build models to forecast product demand using customer information. These models are often integrated with supplier and inventory information to ensure an efficient production operation that meets customer needs. Customer information is also integrated with manufacturing data to model different configurations in order to maximize sales at the same time as controlling configuration costs. The service records associated with products sold are also data mined to detect causes of any malfunctions as well as aiding in the planning of maintenance schedules.

5.3.6   Entertainment

The entertainment and media industry encompasses many diverse organizations. The generation and distribution of content from films, television and radio programs, music, magazines and books comprise a large percentage of the market. In addition, video gaming, sports and other events, hotels and resorts, and casinos are major players in this industry. This industry is increasingly relying on data mining approaches to maximize its operational efficiency, open new markets, and maximize the customer experience.

- *Optimization in entertainment*: Information is critical to the allocation of resources in the entertainment industry. For example, many sports teams are utilizing data concerning player and team performances, strategies, and salary costs. This data is being used to optimize the make-up of individual teams, ensuring the best player combinations at an optimal cost. At the same time, individual game tactics can emerge from an analysis of the data, and models are even built to minimize athletes' injuries. In another example, hotels and resorts gather extensive information for analysis, generating models that help to predict the most profitable daily room price by optimizing occupancy. These models often take into account both historical purchasing information and data concerning the local market.

- *Maximizing advertisement effectiveness*: Many factors contribute to the effectiveness of an advertisement campaign. Data mining approaches are being used to ensure the campaigns are profitable. These approaches make use of historical data to better predict the most effective media to utilize, such as television, radio, print, internet, and so on. Campaigns are also targeted toward specific types of programming, as well as particular days, timeframes, or numbers of advertisements that result in the most effective campaign. Historical data and survey results are often data mined to find unexpected relationships between buying habits and different products. These associations are also used to prioritize advertisement placement. Online advertising now plays an important role and decisions regarding such advertising are often made after making extensive use of analytics. For example, the placement of the online advertisements, or which advertisements are presented in relation to a specific search query, is often directed based on an analysis of usage patterns, which are tracked and logged over time. The collection of detailed information on the behavior of individual internet users has enabled a high degree of advertisement customization, which would not be possible without this tracking.

- *Media*: Online content is often organized and customized through data mining approaches. For example, when newswires are published, automated analysis of these releases is used to categorize the contents, enabling rapid dissemination to interested parties. The content of websites make use of data mining approaches to maximize the delivery of the information, matching specific content or services to the specific user, based on models built from historical usage patterns.

5.3.7 Government

National and local governments have a broad series of responsibility, including providing national security, supporting basic research and development, collecting taxes, and ensuring local resources are available, such as fire prevention services. Data mining approaches are used in many situations within government departments to ensure efficient allocation of resources, to support basic research, and to help detect criminal activities.

- *Revenue collection and resource allocation*: A significant amount of money is wasted as a result of fraud related to either nonpayment of taxes or payment of illegitimate benefits. Analysis of historical data is performed to identify rules associated with this fraudulent activity in order to identify and prevent future occurrences.

- *Planning future needs*: Data mining is also used to provide enhanced levels of public service through more efficient resource allocation. For example, fire departments are using historical data concerning fire emergencies, such as when the incident occurred, how many casualties are involved, as well as the extent and type of damage. This information is used to plan fire station locations, develop proactive fire prevention programs, and optimize allocation of equipment and personnel. Another example is where urban planning departments make use of historical data to plan future population needs, such as transportation needs.

- *Fighting crime*: Solving crimes focuses on the assimilation and interpretation of data from a variety of sources. This may include documents seized, information from witnesses, surveillance data, data on the weather, census data, internet records, email messages, and telephone records. Pulling all this information together is extraordinarily challenging; however, once it is all in a structured electronic format, it can be mined. Understanding associations between specific events or types of events happening in time and over specific geographical areas can play an important role in solving crimes. Information on crimes and arrests within the court system is also collected and is increasingly stored electronically. Mining this information can help in understanding how to balance resources within the court system as well as providing information to lawmakers that could lead to enhanced operational efficiency of the legal system. Information on historical crimes can also aid in the allocation of resources, such as targeting police presence, focusing on areas with the most impact.

- *Research activities*: The government funds a significant portion of basic research within government agencies, universities, and the private sector. Data mining is being used, for example, in the discovery of new materials, identifying the biological basis for disease, and understanding astronomical databases.

5.3.8 Pharmaceuticals

The development of a new drug is a long and expensive process, often taking around 14 years to identify and bring to market a single novel pharmaceutical. Throughout the research and development (R&D) process, data is collected and interpreted to make decisions about the best way to move forward. The initial investigation revolves around understanding the biological basis for a particular disease. Once the biological basis, or target, is identified, many chemicals are then tested to see if they show any promise of intervening in the selected disease process. This allows the pharmaceutical researcher to identify the types of chemicals that are most promising at this stage. Many possible variations for a particular type of chemical are tested to understand whether they increase the potency, as well as other desirable properties of a drug. These chemicals are also tested for undesirable properties, such as potential side-effects or safety issues, and the most promising chemicals are taken forward and tested further. At this time, a clinical trial is started to understand the dosage, efficacy and safety of the potential drug within increasingly large number of patients. Once a drug has been approved based on the results of the clinical trials, it is continually monitored in the market.

The pharmaceutical research process has changed dramatically in recent years through the introduction of highly automated robotic testing methods that are able to perform experiments on a small scale, and in some cases even on microchips. This industrialization of the research process has resulted in a dramatic increase in the volume of data generated at all stages of the process and the extensive use of data mining is relied upon to make decisions. The volume of data, alongside the need to handle nontraditional data, such as genes, proteins and chemicals, which is not easily incorporated into a relational table, makes data mining pharmaceutical information challenging.

- *Mining biosequence data*: Understanding the biology of diseases is the first step in the process of discovering a new medicine. It is critical to understand the molecular basis of the disease, that is the genes, proteins and cells involved in the disease processes. Automated methods of DNA sequencing, including human and other organisms, have created large gene databases. These databases contain a rich knowledge base for understanding diseases. Genes and proteins are usually represented as sequences of letters that comprise their building blocks. For gene-related sequences there are four letters (representing the bases that make up genes) and for proteins there are 20 letters (the amino acids that make up proteins). These sequences have no predetermined length and so are usually represented as a single string of characters. Searching and grouping similar sequences are extensively used. Unlike more traditional data methods for assessing similarity based on predetermined columns of data, individual sequences must be aligned and compared to each other in order to cluster the sequences. Methods such as clustering and visualization of the data are used to make sense of the vast volumes of data and enlighten the pharmaceutical scientist concerning the biological basis for the disease. As a result of the number of specialized methods necessary to analyze and mine biological data, the field of *bioinformatics* was established. Section 5.4 provides a case study of mining microRNA data.

- *Finding patterns in chemical data*: With an understanding of the biological basis of a particular disease, it is possible to generate an experimental test that, when used in conjunction with a candidate drug (chemical), will determine whether there is a chance this candidate would cure the disease. Any positive results are a very long way off from becoming a drug, but it is a starting point. It is usual to test a large number of chemicals at this point, often hundreds of thousands. This data is then mined to determine what types of chemicals should be further refined in the long process of identifying a drug. The test results can be represented as numeric values, which are cleaned and incorporated as variables into data tables. However, the primary goal at this stage is to relate attributes of the chemicals to these biological test results, and so it is essential to generate independent variables related to the chemicals. There are very many independent variables that describe whole chemical attributes and pieces of the chemical. Once these independent variables have been computed, the chemical and biological data is usually mined by methods such as decision trees and clustering to determine the attributes of those chemicals that correspond to promising biological test results. These attributes are used to design and select the next batch of chemicals to test as the drug discovery process advances. There are many unique issues associated with mining chemical data, such as representation, normalization, cleaning, and independent variable generation, and all of these issues are encompassed into the field of *chemoinformatics*. An overview of how this non tabular data is processed is provided in Section 5.6.2.

- *Optimizing drugs*: Based on the desired chemical attributes identified earlier, sets of chemicals are iteratively tested, meaning that the test results are mined and the results guide the selection of the next batch of chemicals to be tested. In addition to the testing for potency, the chemicals are also tested for other properties, such as to determine how easily the drug will pass through the lining of the stomach and into the blood. A highly potent chemical that cannot get into the blood will never become a viable drug. There are many different attributes that are tested for, commonly referred to as ADME (absorption, distribution, metabolism, and excretion). The process of understanding this data is a complex multidimensional optimization problem and often incorporates extensive predictive analytics.

- *Assessing clinical data*: Having selected the most promising drug to take forward, conducting a series of human clinical trials is usually the last step prior to taking a new drug to market. An initial assessment of the product safety is made in healthy adults, along with a determination of the dosage level. Once this first step is successfully completed, two controlled double blind studies are performed in the clinic. The first study is performed on a smaller number of patients, and then the drug is tested in a wider population if the results from the first study are acceptable. The patient populations are divided into two groups: one group is administered the drug and the other group is administered a placebo. The analysis of the data will answer the question of whether the drug is both effective and safe. Once a drug is marketed, the drug is continually monitored for undesirable properties, and the drug may be withdrawn if any safety issues come to light. Withdrawing a drug from the market as a result of safety concerns, however, can be disastrous for both the patient's health and the financial viability of the pharmaceutical company.

- *Predicting safety issues*: Throughout the entire process of drug discovery, large collections of data are generated concerning both the desirable and undesirable properties of drugs or drug candidates. This information is gathered from within an organization and from published information. The process of drug discovery is accelerated by using predictive models to complement or as an alternative to physical safety testing. These models are used to prioritize research directions and to avoid taking drug candidates with potential problems further. Collecting and normalizing the data is challenging since the chemicals may have been tested using different types of experiments or experimental protocols. The results are often collected from controlled experiments, generated for a specific type of chemicals. To use this data to make predictions concerning the general population of possible chemicals requires care in putting the training sets together. This training set should ideally now represent a diverse set of chemicals to increase the applicability of any predictive model generated. In reality, the types of chemicals in the training sets limit what types of chemicals can be provided as input to models. It is usual to make an assessment of whether a particular chemical can be used with a particular model by comparing the chemical to be tested against the training set of the model. When the chemical to be predicted is outside this *applicability domain*, a prediction would not be reliable.

5.3.9 Healthcare

The healthcare industry covers organizations involved in the diagnosis, treatment, and care of patients. Hospitals, care facilities, physicians, and dentists are the primary providers of healthcare, with support from industries such as pharmacies and medical and diagnostic laboratories. Hospitals and physician offices are starting to use electronic health or medical record systems where data is collected concerning the patient, including clinical notes, prescriptions, and laboratory results. Hospitals also collect information on patient workflows and available resources, as well as other data that may affect the day-to-day operations of a hospital such as weather-related data or information on particular community events. Data mining is being extensively used to accelerate the pace of research and improve diagnosis, treatment, and care for patients.

- *Improving care*: Computer systems are being put in physicians' offices and hospitals to record patient data, such as clinical notes, prescriptions and laboratory results. These electronic health or medical record systems (EHR/EMR) are storing information on a patient's medical history, including symptoms, diagnosis, treatments, and results. This information is often combined with information on demographic and geographical patient data, along with the hospital resources, such as staff, beds, equipment, and operating theaters. The data is mined to optimize care and help with future diagnosis.

- *Assessing treatment outcomes*: Data that has been collected across different patients, doctors, and facilities can be used to make assessments concerning the effectiveness of care for various outcomes. For example, an outcome may be patient mortality rates after treatment for a heart attack. This information can be used to judge the effectiveness of differing diagnoses, treatment, and care options, as well as an individual doctor's performance. Data mining this information can help to increase the quality of care by establishing medical practices that result in improved care, such as decreasing mortality rates. When preparing data for this purpose, care should be taken to ensure any bias in patient populations within specific facilities is taken into account by normalizing the data based on a model of patient risk. Models can also be built from this data that take into account information on the patients, such as demographic data, to make predictions of the type of care that leads to the best outcome for an individual patient.

- *Improving diagnosis*: Early diagnosis of problems or identification of potential long-term side effects can positively impact the patient and can result in significantly lower costs. Methods such as decision trees are often built from the historical clinical data, including the patient's symptoms, health history, and demographics, to predict future issues, and provide an opportunity for intervention. Data collected is often noisy, with many missing values, and needs to be cleaned before building any models. This cleaning may also need to take into account standardization of the vocabulary used and the incorporation of concept hierarchies to map many terms into a more standardized list. Coupling historical diagnosis data with information generated at a genetic level, such as gene expression data, is now starting to be used to enhance diagnosis and enable more personalized medicine.

- *Allocating resources*: Data mining is also being used to optimize the operational efficiency of hospitals. Models are built to make predictions of demand on resources such as hospital beds, theaters, and staff. These models help predict demand more accurately on a day-to-day basis.

5.4   microRNA DATA ANALYSIS CASE STUDY

5.4.1   Defining the Problem

To gain insight into the molecular basis for cancer, the National Cancer Institute (NCI), along with other institutions, has run extensive experiments on 60 human cancer cell lines. These cell lines are summarized in Table 5.1, and are organized by the following tissue types: breast, central nervous system (CNS), colon, lung, leukemia, melanoma, ovarian, prostate, and renal. A variety of tests have been performed on these cell lines over the years, including chemical screening and profiling based on gene and protein expression levels. The NCI has built up an extensive collection of experimental research data based on these cell lines and the associated experiments (http://discover.nci.nih.gov/cellminer/).

microRNAs are thought to play a critical role in cancer, and recently scientists at the Ohio State University have performed a series of tests to profile the 60 cancer cell lines by microRNA expression levels using a custom microarray (Blower et al., 2007). These expression levels indicate the relative abundance of each microRNA in each of the cell lines. A total of 279 microRNAs were tested against the 60 cancer cell lines detailed in Table 5.1.

The data has been generated to answer a number of questions, including:

1. Is it possible to use the data to classify tissue types, such as melanoma, lung, and so on, by microRNA expression patterns?

2. Is it possible to predict the tissue type by microRNA expression level?

5.4.2   Preparing the Data

An extensive amount of transformation of the data has been performed, which is detailed in Blower et al. (2007). A portion of the resulting data table is shown in Table 5.2. Each row corresponds to a specific microRNA (such as let-7a-1 or let-7a-2-precNo2), and each column is a specific cancer cell line (*BR:BT-549*,

TABLE 5.1   The 60 Human Cancer Cell Lines Organized by Tissue Type

Tissue	Cell names
Breast	BT-549, HS578T, MCF7, MDA-MB-231, T47D
CNS	SF.268, SF.295, SF.539, SNB19, SNB.75, U251
Colon	COLO205, HCC-2998, HCT.116, HCT.15, HT29, KM12, SW-620
Lung	A549-ATCC, EKVX, HOP-62, HOP-92, NCI-H226, NCI-H23, NCI-H322M, NCI-H460, NCI-H522
Leukemia	CCRF-CEM, HL-60, K-562, MOLT-4, RPMI-8226, SR
Melanoma	LOXIMVI, M14, MALME-3M, MDA-MB-435, MDA-N, SK-MEL-2, SK-MEL-28, SK-MEL-5, UACC-257, UACC-62
Ovarian	IGROV1, NCI-ADR-RES, OVCAR-3, OVCAR-4, OVCAR-5, OVCAR-8, SKOV3
Prostate	DU-145, PC-3
Renal	786-0, A498, ACHN, CAKI-1, RXF-393, SN12C, TK-10, UO-31

TABLE 5.2 Table Showing the Expression Level of microRNAs for the 60 Human Cancer Cell Lines

	BR:BT-549	BR:HS578T	BR:MCF7	BR:MDA-MB-231	BR:T47D	CNS:SF-268	CNS:SF-295	CNS:SF-539	CNS:SNB-19	CNS:SNB-75	
let-7a-1-prec	12.39	12.42	11.32	12.18	10.62	11.91	11.94	11.5	12.4	13.16	...
let-7a-2-precNo2	12.66	13.15	12.05	12.28	10.55	13.11	12.86	12.1	13.65	13.65	...
let-7a-3-prec	11.76	12.34	11.05	11.6	9.99	12.09	11.98	11.08	12.63	12.92	...
let-7b-prec	12.62	13.28	11.75	12.58	10.71	12.87	12.46	11.8	13.36	13.58	...
let-7c-prec	12.56	13.42	11.96	12.08	10.49	12.92	12.85	11.95	13.85	13.7	...
let-7d-prec	11.63	12.28	11.16	10.94	9.17	11.93	11.94	11.01	12.83	12.65	...
let-7d-v1-prec	10.41	10.95	9.98	10.12	8.07	10.49	10.32	9.91	10.85	11.43	...
let-7d-v2-precNo2	11.16	11.99	10.74	10.59	8.14	10.8	10.6	10.51	11.13	11.6	...
let-7e-prec	11.35	11.74	11.13	11.43	9.79	11.46	11.25	10.99	11.7	11.93	...
let-7f-1-precNo2	9.92	9.35	9.48	10.02	7.55	10.31	9.3	8.97	9.84	10.62	...
let-7f-2-prec2	10.55	10.16	10.32	10.16	7.94	10.71	10.15	10.04	10.71	11.58	...
let-7g-precNo1	11.11	11.06	10.53	10.82	8.26	10.57	10.31	10.67	11.02	11.7	...
...	...	...	...	...	...	...	...	...	...	...	...

TABLE 5.3 Expression Data with Cell Lines as Rows and the microRNAs as the Columns

	Type	let-7a-1-prec	let-7a-2-precNo2	let-7a-3-prec	let-7b-prec	let-7c-prec	let-7d-prec	...
BR:BT-549	Breast	12.39	12.66	11.76	12.62	12.56	11.63	...
BR:HS578T	Breast	12.42	13.15	12.34	13.28	13.42	12.28	...
BR:MCF7	Breast	11.32	12.05	11.05	11.75	11.96	11.16	...
BR:MDA-MB-231	Breast	12.18	12.28	11.6	12.58	12.08	10.94	...
BR:T47D	Breast	10.62	10.55	9.99	10.71	10.49	9.17	...
CNS:SF-268	CNS	11.91	13.11	12.09	12.87	12.92	11.93	...
CNS:SF-295	CNS	11.94	12.86	11.98	12.46	12.85	11.94	...
CNS:SF-539	CNS	11.5	12.1	11.08	11.8	11.95	11.01	...
CNS:SNB-19	CNS	12.4	13.65	12.63	13.36	13.85	12.83	...
CNS:SNB-75	CNS	13.16	13.65	12.92	13.58	13.7	12.65	...
CNS:U251	CNS	12.16	12.6	11.59	12.89	12.51	11.62	...
CO:COLO205	Colon	10.49	11.95	10.98	11.57	11.85	11.01	...
CO:HCC-2998	Colon	11.19	11.68	10.77	11.89	11.66	10.64	...
CO:HCT-116	Colon	11.36	12.78	11.87	11.95	12.72	11.78	...
...	...	...	...	...	...	...	...	...

BR:HS578T, and so on, where the prefix indicates tissue type). The data values are the expression level of the microRNA for each of the 60 cell lines. It should be noted that the data values have been *log2* transformed.

The rows and columns are initially exchanged (matrix transpose), such that the rows are now the 60 cancer cell lines, and the columns are the microRNA probe IDs, as shown in Table 5.3.

Each cell line is classified according to its tissue type using Table 5.1 and this process generates a new variable (*type*). This variable is also transformed into a series of dummy variables, one for each tissue: breast, CNS, colon, and so on, as shown in Table 5.4.

5.4.3 Analysis

The data in Table 5.4 is initially clustered using the microRNA expression patterns. The clustering is performed to determine whether the cell lines are grouped by tissue type. Hierarchical agglomerative clustering using the complete linkage joining rule was performed. This clustering used the correlation coefficients of the microRNA expression levels. The results are shown in Fig. 5.1.

The dendrogram in Fig. 5.1 is annotated with the individual cell line names. The prefix of the name indicates its tissue type, for example, LE:K-582 is from a leukemia tissue, and CO:SW-620 is from a colon tissue. The dendrogram indicates that many of the cell lines are effectively grouped together, in particular leukemia (LE), colon (CO), renal (RE), melanoma (ME), and CNS. Two exceptions are the breast and lung cancer cell lines.

TABLE 5.4 A Series of Dummy Variables Generated for Each Tissue Type

	Type	let-7a-1-prec	...	mir_320_Hcd306 right	Type								
					Breast	CNS	Colon	Lung	Leukemia	Melanoma	Ovarian	Prostate	Renal
BR:BT-549	Breast	12.39	...	12.7	1	0	0	0	0	0	0	0	0
BR:HS578T	Breast	12.42	...	11.89	1	0	0	0	0	0	0	0	0
BR:MCF7	Breast	11.32	...	10.8	1	0	0	0	0	0	0	0	0
BR:MDA-MB-231	Breast	12.18	...	11.46	1	0	0	0	0	0	0	0	0
BR:T47D	Breast	10.62	...	11.2	1	0	0	0	0	0	0	0	0
CNS:SF-268	CNS	11.91	...	12.18	0	1	0	0	0	0	0	0	0
CNS:SF-295	CNS	11.94	...	11.21	0	1	0	0	0	0	0	0	0
CNS:SF-539	CNS	11.5	...	11.93	0	1	0	0	0	0	0	0	0
CNS:SNB-19	CNS	12.4	...	11.15	0	1	0	0	0	0	0	0	0
CNS:SNB-75	CNS	13.16	...	12	0	1	0	0	0	0	0	0	0
CNS:U251	CNS	12.16	...	11.27	0	1	0	0	0	0	0	0	0
CO:COLO205	Colon	10.49	...	10.83	0	0	1	0	0	0	0	0	0
CO:HCC-2998	Colon	11.19	...	11	0	0	1	0	0	0	0	0	0
CO:HCT-116	Colon	11.36	...	11.71	0	0	1	0	0	0	0	0	0
...	...	...	...	...	...	...	...	...	...	...	...	...	...

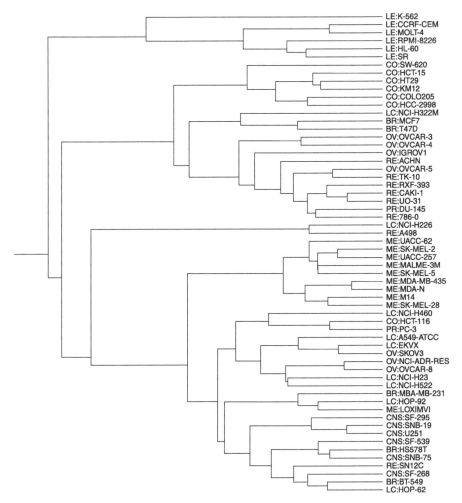

Figure 5.1   Clustering dendrogram of the human cancer cell lines by microRNA expression patterns

The data from Table 5.4 is additionally grouped using a decision tree. The microRNA expression levels (*let-7a-1-prec*, *let-7a-2-precNo*, and so on) are used as independent variables. The *type* variable, whose values are the nine tissue types, that is breast (BR), colon (CO), and so on, is used as the response variable. A decision tree was generated using these variables, where the generation process is restricted to nodes with more than four members. The results are shown in Fig. 5.2.

Each of the nodes in the tree represents a set of cell lines. Starting with the node at the top of the tree containing all observations, the tree is continuously split. These splits attempt to generate nodes ideally containing a single type of cell line. The criteria used to split each node are shown above the node. For example, the initial split divides all observations into those where *mir-200cNo1* < 6.87, and those where

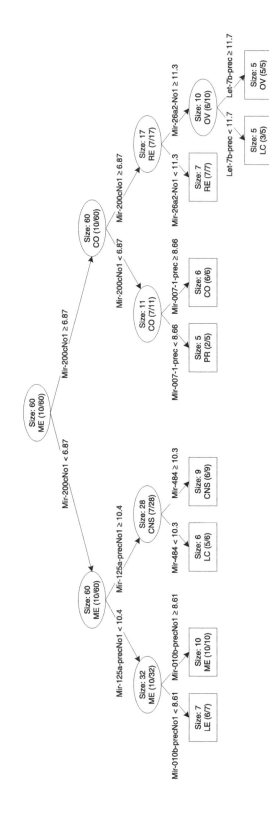

Figure 5.2 Decision tree generated for classifying microRNA expression level according to tissue type

mir-200cNo ≥ 6.87. The observations are continually split until either the minimum node threshold has been reached or all the observations are assigned to a single class. Nodes at the end of the branches are called terminal nodes, and the total number of observations in each node is shown, as well as the number of observations in the most common class.

The criteria used to generate each terminal node can be defined by tracing the decision points back to the original node. For example, the node that classifies the 10 melanoma cell lines is defined by the rule: *mir-200cNo1* < 6.87 AND *mir-125a-precNo1* < 10.4 AND *mir-010b-precNo* ≥ 8.61.

Many of the terminal nodes characterize cell lines according to their tissues types. The tree is able to classify LE, ME, CNS, ovarian (OV), CO, and RE cell lines well.

The second question to explore was whether a model could be generated to predict the tissue type by microRNA expression level. Melanoma tissues will be used to test this, using Table 5.4 (*type = melanoma*) containing the microRNA expression levels (*let-7a-1-prec*, *let-71-2-precNo*, and so on). The data set in Table 5.4 contains 60 observations with 279 independent variables. The first step is to identify a subset of variables to use in any model. There are many approaches to reducing the number of variables prior to modeling, including principal component analysis, decision trees, as well as clustering the variables and selecting a representative variable from each cluster.

For this exercise, a hypothesis test was performed on each variable comparing two groups of observations, those where the variable *type = melanoma* is 1 and those where *type = melanoma* is 0. Based on the mean of the two groups, a *t*-test is generated which can be used to prioritize the variables. Figure 5.3 is an example of a hypothesis test performed on one of the variables (*mir-509No1*).

A table is generated summarizing the results for each variable. The mean for both of the two groups is calculated where melanoma is 1 (*M* = 1) and where melanoma is 0 (*M* = 0), along with the *t*-statistic, which is used to prioritize the list. The top 15 variables are shown in Table 5.5.

Assessing variable:	mir-509No1	
	Group 1	Group 2
Observations:	Where type = melanoma is 1	Where type = melanoma is 0
Count:	10	50
Mean:	7.185	2.355
Null hypothesis:	Group 1 mean = Group 2 mean	
Alternative hypothesis:	Group 1 mean ≠ Group 2 mean	
Alpha (α):	0.1	
Hypothesis score (*t*-):	12.439	
Critical value (*t*-):	1.671	
p-value:	<0.005	
Conclusion:	Reject the null hypothesis	

Figure 5.3   Hypothesis test performed on the variable mir-509No1 for two groups

TABLE 5.5 Top 15 Variables Prioritized by the *t*-Test Scores

		Mean ($M = 1$)	Mean ($M = 0$)	Hypothesis (t-)
1	mir-509No1	7.2	2.4	12.4
2	mir-146-prec	14.4	8.5	8.3
3	mir-146bNo1	13.9	8.3	7.7
4	mir-204-precNo1	9.2	3.9	6.7
5	mir-211-precNo1	9.082	5	6.4
6	mir-147-prec	8.6	5.8	6.4
7	mir-182-precNo2	6.5	9.3	4.7
8	mir-335No1	2.3	5.5	4
9	mir-200cNo1	3	7.3	3.8
10	mir-224-prec	2.6	6	3.6
11	mir-183-precNo1	4.6	6.9	3.6
12	mir-191-prec	11	12	3.4
13	mir-200bNo1	4.5	8.5	3.4
14	mir-335No2	8.6	9.8	3.3
15	mir-200a-prec	3.4	7.4	3.2
...	...	...	...	...

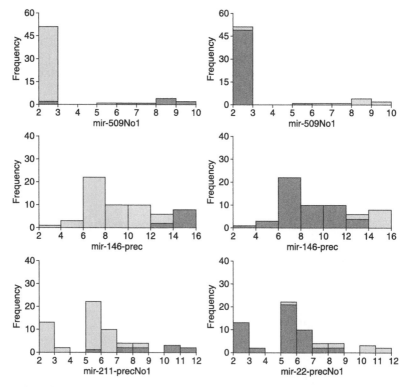

Figure 5.4   Comparison of three prioritized variables where the highlighted observations in the first column are for melanoma $= 1$ whereas the highlighted observations in the graphs in the second column are for melanoma $= 0$.

	mir-146-prec	mir-146bNo1	mir-147-prec	mir-182-precNo2	mir-183-precNo1	mir-191-prec	mir-200a-prec	mir-200bNo1	mir-200cNo1	mir-204-prceNo1	mir-211-prceNo1	mir-224-prce	mir-335No1	mir-335No2	mir-509No1
mir-146-prec	1	0.965	0.473	0.201	0.124	0.0767	0.112	0.076	0.161	0.191	0.19	0.0128	0.224	0.16	0.392
mir-146bNo1	0.965	1	0.446	0.206	0.117	0.0802	0.131	0.0803	0.18	0.163	0.168	0.017	0.235	0.148	0.373
mir-147-prec	0.473	0.446	1	0.222	0.114	0.106	0.0518	0.0448	0.0734	0.244	0.375	4.95E-5	0.0991	0.0495	0.315
mir-182-precNo2	0.201	0.206	0.222	1	0.711	0.0696	0.258	0.248	0.145	0.0706	0.19	0.00355	0.166	0.0693	0.261
mir-183-precNo1	0.124	0.117	0.114	0.711	1	0.0581	0.285	0.384	0.172	0.0274	0.127	7.92E-6	0.0765	0.045	0.136
mir-191-prec	0.0767	0.0802	0.106	0.0696	0.0581	1	0.0223	0.0794	0.0548	0.0617	0.109	0.12	0.146	0.0622	0.103
mir-200a-prec	0.112	0.131	0.0518	0.258	0.285	0.0223	1	0.747	0.85	0.127	0.151	0.00712	0.0917	0.0976	0.139
mir-200bNo1	0.076	0.0803	0.0448	0.248	0.384	0.0794	0.747	1	0.669	0.0849	0.171	0.0518	0.0449	0.0531	0.105
mir-200cNo1	0.161	0.18	0.0734	0.145	0.172	0.0548	0.85	0.669	1	0.158	0.161	0.0369	0.0878	0.11	0.17
mir-204-precNo1	0.191	0.163	0.244	0.0706	0.0274	0.0617	0.127	0.0849	0.158	1	0.541	0.0418	0.0263	0.0182	0.442
mir-211-precNo1	0.19	0.168	0.375	0.19	0.127	0.109	0.151	0.171	0.161	0.541	1	0.0347	0.106	0.0691	0.36
mir-224-prec	0.0128	0.017	4.95E-5	0.00355	7.92E-6	0.12	0.00712	0.0518	0.0369	0.0418	0.0347	1	0.0743	0.048	0.143
mir-335No1	0.224	0.235	0.0991	0.166	0.0765	0.146	0.0917	0.0449	0.0878	0.0263	0.106	0.0743	1	0.809	0.138
mir-335No2	0.16	0.148	0.0495	0.0693	0.045	0.0622	0.0976	0.0531	0.11	0.0182	0.0691	0.048	0.809	1	0.0728
mir-509No1	0.392	0.373	1.315	0.261	0.136	0.103	0.139	0.105	0.17	0.442	0.36	0.143	0.138	0.0728	1

Figure 5.5 Squared correlation coefficients for the 15 prioritized variables

189

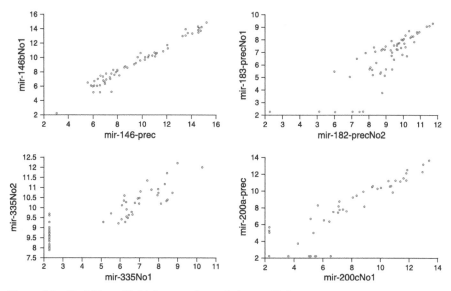

Figure 5.6   Variables with high squared correlation coefficients

Three of these variables from this list (*mir-509No1*, *mir-146-prec*, and *mir-211-precNo1*) are shown in Fig. 5.4. The frequency distributions on the left highlight the observations where the tissue type is melanoma, and on the right they highlight where the tissue type is not melanoma. There is a clear differentiation of the melanoma tissue type based on these variables.

These prioritized variables are further investigated to determine whether there are relationships between the variables. The existence of a relationship between independent variables would be redundant, and this colinearity could be problematic to the model building process. A matrix of the squared correlation coefficients is generated and shown in Fig. 5.5. It is determined that there exists a certain amount

TABLE 5.6   Prioritized Variables with Highly Correlated Variables Removed

		Mean ($M = 1$)	Mean ($M = 0$)	Hypothesis (z-)
1	mir-509No1	7.2	2.4	12.4
2	mir-146-prec	14.4	8.5	8.3
4	mir-204-precNo1	9.2	3.9	6.7
5	mir-211-precNo1	9.082	5	6.4
6	mir-147-prec	8.6	5.8	6.4
7	mir-182-precNo2	6.5	9.3	4.7
8	mir-335No1	2.3	5.5	4
9	mir-200cNo1	3	7.3	3.8
10	mir-224-prec	2.6	6	3.6
12	mir-191-prec	11	12	3.4
13	mir-200bNo1	4.5	8.5	3.4

TABLE 5.7   A Sample of the Logistic Regression Models Built

Accuracy	Variable (1)	Variable (2)	Variable (3)	Variable (4)	Variable (5)	...
0.95	mir-146-prec					
0.917	mir-147-prec					
0.833	mir-182-precNo2					
0.833	mir-200cNo1					
0.917	mir-204-precNo1					
0.9	mir-211-precNo1					
0.833	mir-335No1					
0.95	mir-509No1					
0.9	mir-146-prec	mir-147-prec				
0.883	mir-146-prec	mir-182-precNo2				
0.883	mir-146-prec	mir-200cNo1				
0.933	mir-146-prec	mir-204-precNo1				
1	mir-146-prec	mir-211-precNo1				
0.85	mir-146-prec	mir-335No1				
...	...	...				
0.95	mir-146-prec	mir-147-prec	mir-182-precNo2			
...	...	...	...			
0.933	mir-146-prec	mir-147-prec	mir-182-precNo2	mir-200cNo1		
...	...	...	...	...		
0.983	mir-146-prec	mir-147-prec	mir-182-precNo2	mir-200cNo1	mir-204-precNo1	
...	...	...	...	...	...	...

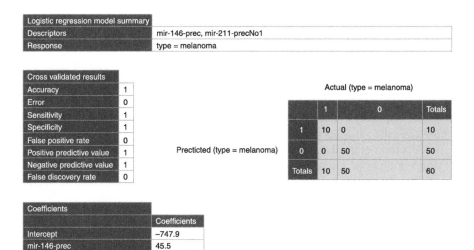

Figure 5.7   Summary of the logistic regression model

of overlap between some of these variables, since some of the squared correlation coefficient values are close to 1.

Four pairs of variables with high squared correlation coefficients are also displayed in Fig. 5.6, to illustrate the relationships between these variables.

The prioritized variables from Table 5.5 are further reduced based on the relationships between them. Where two variables have a high square correlation coefficient score, the lower ranking variable is removed. The resulting table is shown in Table 5.6.

Using the top eight variables from Table 5.6, a series of logistic regression models were built using these variables as independent variables. The response variable is the binary variable *type = melanoma*, and a 5% cross-validation was performed. All combinations of these eight variables were used as inputs to the models, generating 255 different models. These models were then ranked on their level of accuracy. A fraction of the 255 models can be seen in Table 5.7.

The simplest model with the highest cross-validated predictive accuracy was selected. This model contains two independent variables: *mir-146-prec* and *mir-211-precNo1*. The full logistic regression model can be seen in Fig. 5.7.

5.5 CREDIT SCORING CASE STUDY

5.5.1 Defining the Problem

A financial services company has collected data concerning automobile applications and loans, as well other information such as their client's credit performance. Over the years, the company has processed many loan applications, and has recorded those where they had to shut down the account as a result of nonpayment. The data shows that 32% of all customers defaulted on their car loans. The company would like to develop a prediction model for screening loan candidates in order to understand those individuals not likely to default on any loan. Even a small increase in the screening of candidates would be beneficial.

5.5.2 Preparing the Data

A data set has been provided by Dr Satish Nargundkar (http://www.nargund.com/gsu/index.htm) which contains 14,042 observations concerning auto loans, credit histories, application data, and information concerning the vehicle. Table 5.8 summarizes the variables contained in the data set.

A portion of the data table is shown in Table 5.9. The table contains a number of different types of variables, such as text, continuous, binary, and so on.

To facilitate the analysis, a glossary of terms commonly used in the field of auto loans has been put together, and is shown in Table 5.10.

Preliminary Assignment of Variables The objective of the modeling exercise is to build a model to predict the variable *good*, which will be used as the response variable. A preliminary inspection of the data indicates that the *bad* variable is the inverse of *good*, and hence it will not be used in any further analysis. The variable *acctno* is a unique number for each row, and will be used as a label for the observations. All other

TABLE 5.8   Description of the Variables in the Credit Scoring Data Set

Variable name	Description
ACCTNO	Account number
AGEAVG	Average age of trades
AGEOTD	Age of oldest trade
AUUTIL	Ratio of balance to highest credit limit for all open auto trades
BAD	Performance (charged off in 12 months)
BKTIME	Time since bankruptcy
BRHS2X	Number of bank revolving trades ever 30 days past due
BRHS3X	Number of bank revolving trades ever 60 days past due
BRHS4X	Number of bank revolving trades ever 90 days past due
BRHS5X	Number of bank revolving trades ever 120+ days past due
BROLDT	Age of oldest bank revolving trade
BROPEN	Number of open bank revolving trades
BRTRDS	Number of bank revolving trades
BSRETL	Base retail value
BSWHOL	Base wholesale value
CBTYPE	Credit bureau type
CFTRDS	Number of financial trades
CONTPR	Contract price paid
CURSAT	Number of trades currently rated satisfactory
DWNPMT	Down payment
GOOD	Performance (not charged off)
HSATRT	Ratio of satisfactory trades to total trades
INQ012	Number of inquiries in last 12 months
MAKE	Make of automobiles
MILEAG	Mileage
MNGPAY	Monthly gross pay
MODEL	Model of automobile
NEWUSE	New or used indicator
NTTRIN	Net trade in
PUBREC	Number of derogatory public records
TERM	Term of loan
TOTBAL	Total balance
TRADES	Number of trades
VAGE	Customer age
VDDASAV	Checking/saving accounts
VJOBMOS	Time at job in months
VRESMOS	Time at residence in months
AGEOTD	Age of oldest trade
BKRETL	Book retail value
BRBAL1	Number of open bank revolving trades with balance $\geq\$1000$
CSORAT	Ratio of currently satisfactory trades to open trades
HST03X	Number of trades never 90 days past due
HST79X	Number of trades ever rated bad debt

(Continued)

TABLE 5.8 *Continued*

Variable name	Description
MODLYR	Automobile model year
OREVTR	Number of open revolving trades
ORVTB0	Number of open revolving trades with balance > $0
REHSAT	Number of retail trades ever rated satisfactory
RVOLDT	Age of oldest revolving trade
RVTRDS	Number of revolving trades
T2924X	Number of trades rated greater than 30 days past due in the last 24 months
T3924X	Number of trades rated greater than 60 days past due in last 24 months
T4924X	Number of trades rated greater than 90 days past due in last 24 months
TIME29	Months since most recent greater than 30 days past due rating
TIME39	Months since most recent greater than 60 days past due rating
TIME49	Months since most recent greater than 90 days past due rating
TROP24	Number of trades opened in last 24 months

TABLE 5.9 Portion of the Data Set Concerning Auto Loan Applications

ACCTNO	AGEAVG	AGEOTD	AUUTIL	BAD	BKTIME	BRHS2X	BRHS3X	...
1004596	−5	−5	−5	0	−6	−5	−5	...
1004598	46	81	−6	1	−6	−6	−6	...
1004599	−2	−2	−2	0	−2	−2	−2	...
1004602	39.52	216	−6	1	−6	0	1	...
1004603	30.44	107	−6	0	36	−6	−6	...
1004610	−3	−3	−3	0	−3	−3	−3	...
1004614	40	62	−6	0	−6	0	0	...
1004615	39.67	79	−6	1	−6	−6	−6	...
1004616	13	69	−6	0	−6	0	1	...
1004619	−2	−2	−2	0	−2	−2	−2	...
1004623	27.6	109	−6	0	−6	1	2	...
1004624	37.83	93	−6	0	−6	−6	−6	...
1004625	46.25	79	−6	1	−1	−6	−6	...
1004626	79.5	246	−6	1	63	0	0	...
1004627	30.56	80	−6	0	−6	0	0	...
1004628	−1	32	−6	1	−6	−6	−6	...
...	...	...	...	...	...	...	...	...

TABLE 5.10 Glossary of Auto Loan Terminology

Term	Description
Trade	A trade is a financial contract, like a Visa card, automobile loans, or a mortgage
Revolving trades	A trade, such as a credit card, with no fixed installments
Days past due	Number of days a payment is late in 30 day increments
DDA	Demand deposit account or basic checking account

variables are candidates to be used as independent variables, to be used in building the models.

Negative Observations   Many fields contain negative values, and these values are used to indicate a specific situation where a value cannot be easily assigned. For example, a variable may indicate the number of trades (see glossary) that an individual has not paid on time. A zero value would indicate that the individual has paid all trades on time; however, a negative value is used to indicate the individual does not have any trades to report. Table 5.11 is an analysis of the variables showing the number of negative values in each of the variables. The table reflects that 1606 individuals have no credit history. Other negative numbers reflect missing data concerning that particular type of trade.

TABLE 5.11   Number of negative observations for each variable

Variable name	Negative observations	Variable name	Negative observations
ACCTNO	0	NTTRIN	0
AGEAVG	2521	PUBREC	1522
AGEOTD	1666	TERM	0
AUUTIL	12491	TOTBAL	1606
BAD	0	TRADES	0
BKTIME	11969	VAGE	0
BRHS2X	7789	VDDASAV	0
BRHS3X	7789	VJOBMOS	0
BRHS4X	7789	VRESMOS	0
BRHS5X	7789	AGEOTD	1666
BROLDT	7795	BKRETL	0
BROPEN	7789	BRBAL1	11182
BRTRDS	1606	CSORAT	6877
BSRETL	0	HST03X	1606
BSWHOL	0	HST79X	1606
CBTYPE	0	MODLYR	0
CFTRDS	1606	OREVTR	5158
CONTPR	0	ORVTB0	9097
CURSAT	1606	REHSAT	7294
DWNPMT	0	RVOLDT	5164
GOOD	0	RVTRDS	1606
HSATRT	1606	T2924X	1606
INQ012	582	T3924X	1606
MAKE	0	T4924X	1606
MILEAG	0	TIME29	2219
MNGPAY	0	TIME39	2372
MODEL	0	TIME49	2473
NEWUSE	0	TROP24	1606

Handling Missing Data The variable *VAGE* contains 40 observations with missing values. These observations are removed.

Transforming Variables There are a number of variables that contain only text values: *make, model, newuse,* and *VDSASAV. Make* and *model* contain many values, and each value has either a small number of observations, or values that do not appear to have a strong relationship to the response variable *good*. These variables are not converted to numbers, and will not be used in any model. The variables *newuse* contains values "*U*" and "*N*," and *VDASAV* contains values "DDA," "none," "both," and "SAV." They are converted to dummy variables and may be used to build models. The following new variables are generated: *newuse = U, newuse = N, VDASAV = DDA, VDASAV = none, VDDASAV = both,* and *VDDASAV = SAV.*

Data Outliers In examining the frequency distribution of all the variables, a number of values that were significantly higher than the majority of cases were identified. Figure 5.8 shows a number of examples where the frequency distribution contained these outliers. These observations were removed from the data set.

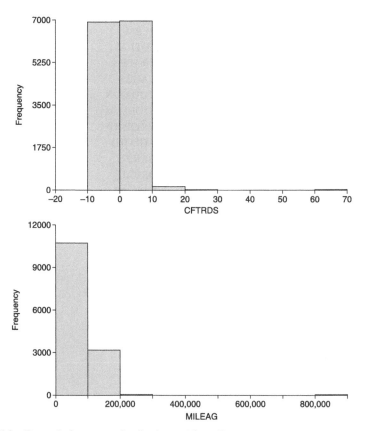

Figure 5.8   Example frequency distributions with outliers

Variance explained:

Principal components	PC1	PC2	PC3	PC4	PC5	PC6	PC7	PC8	PC9	PC10
Eigenvalues	12.8	5.61	5.26	3.8	2.24	2.01	1.88	1.76	1.64	1.37
Percentage	33.4%	14.6%	13.7%	9.9%	5.82%	5.24%	4.89%	4.57%	4.27%	3.58%

Loadings:

	PC1	PC2	PC3	PC4	PC5	PC6	PC7	PC8	PC9	PC10
ACCTNO	0.0161	−0.062	−0.0252	−0.0162	−0.0284	−0.0477	−0.181	−0.125	−0.0401	0.0171
AGEAVG	0.153	0.0337	−0.118	−0.236	0.0496	0.0194	−0.00655	−0.0559	0.283	0.105
AGEOTD	0.193	0.0622	−0.116	−0.203	0.0352	0.0245	0.0115	−0.059	0.315	0.0296
AUUTIL	0.0574	0.0285	0.00648	0.0505	−0.141	−0.0155	0.0709	0.0504	−8.61E-4	−0.309
BAD	−0.0384	0.116	0.0148	0.107	0.223	0.062	0.533	0.295	0.113	0.0508
BKTIME	0.0587	0.0104	−0.0395	−0.0818	0.0672	0.00242	−0.05	−0.0172	0.139	−0.175
BRHS2X	0.193	−0.0876	0.242	0.0193	0.23	−0.00496	−0.0434	−0.0138	−0.0666	−0.0614
BRHS3X	0.193	−0.0869	0.242	0.0219	0.236	−0.00448	−0.0474	−0.0153	−0.0651	−0.0609
BRHS4X	0.191	−0.0864	0.243	0.0217	0.239	−0.00332	−0.0476	−0.0164	−0.0632	−0.06
BRHS5X	0.193	−0.0849	0.24	0.0224	0.238	−0.00376	−0.0479	−0.0199	−0.0581	−0.0594
BROLDT	0.209	−0.0242	0.072	−0.105	0.144	0.00896	−0.0325	−0.0452	0.162	0.0183
BROPEN	0.204	−0.0861	0.241	0.0132	0.183	−0.00712	−0.0234	−0.00167	−0.0978	−0.0587
BRTRDS	0.229	0.0469	−0.00445	−0.02	0.00911	0.00209	0.0029	0.0143	−0.1	−0.00772
BSRETL	0.0633	−0.323	−0.183	0.0964	0.00996	−0.00636	0.0692	0.0101	−1.06E-4	−0.0656
BSWHOL	0.063	−0.324	−0.184	0.0972	0.00953	−0.00367	0.0683	0.0151	0.00194	−0.0583
CBTYPE	−0.115	−0.133	0.231	0.0361	0.0208	0.0021	−0.0106	−0.039	0.17	0.0267
CFTRDS	0.162	0.108	−0.103	0.0163	−0.133	−0.0097	0.0351	−0.00111	−0.0295	−0.156
CONTPR	0.0601	−0.332	−0.192	0.0986	0.0415	−0.0125	0.0577	0.0194	0.0149	0.0185
CURSAT	0.209	0.0519	−0.0518	−0.0462	−0.245	−0.0128	0.112	0.0599	−0.176	0.0141
DWNPMT	0.0334	−0.215	−0.122	0.0809	0.03	−0.0574	0.0634	−0.0488	0.0485	−0.339
GOOD	0.0384	−0.116	−0.0148	−0.107	−0.223	−0.062	−0.533	−0.295	−0.113	−0.0508
HSATRT	0.138	−0.0141	0.0286	−0.0732	−0.334	0.0063	0.126	0.0533	−0.0969	−0.0351
INQ012	0.0676	0.0178	−0.0415	0.0914	−0.0481	−0.0518	0.0873	0.0495	−0.0703	−0.0758
MILEAG	−0.039	0.187	0.0901	−0.0352	−0.0195	−0.0293	0.0114	−0.0126	−0.0228	−0.203
MNGPAY	0.0175	−0.0235	−0.0231	−0.014	−0.0316	−0.00261	0.0211	0.0316	−0.0114	−0.105
NTTRIN	0.0215	−0.0382	−0.0346	−0.0171	0.00642	0.0294	−0.0588	0.0124	−0.0692	0.425
PUBREC	0.103	0.1	−0.185	−0.045	0.0656	0.0283	−0.0295	0.0101	0.0246	−0.174
TERM	0.0471	−0.291	−0.166	0.083	0.0286	0.0149	0.0464	−0.00814	0.00545	0.137
TOTBAL	0.138	0.0257	−0.0129	0.0188	−0.0955	0.00492	0.0176	0.0234	0.111	−0.205
TRADES	0.217	0.119	−0.0727	0.133	−0.00924	−0.00821	−0.0363	−0.0325	0.0457	0.0231
VJOBMOS	0.0701	−0.0153	−0.0274	−0.0792	−0.0561	−0.0151	0.0349	−0.00501	0.116	−0.341
VRESMOS	0.0193	−0.0205	−2.26E-4	−0.0484	−0.0386	0.0131	0.0271	−0.0768	0.117	−0.239
AGEOTD	0.193	0.0622	−0.116	−0.203	0.0352	0.0245	0.0115	−0.059	0.315	0.0296
BKRETL	0.0674	−0.336	−0.19	−0.0959	0.0172	−0.00934	0.0604	0.0107	0.00943	−0.0242
BRBAL1	0.119	−0.105	0.233	0.0105	−0.0605	−0.0111	0.06	0.0299	−0.127	−0.022
CSORAT	0.157	−0.022	0.0433	−0.0583	−0.252	−0.00446	0.125	0.046	−0.089	0.0523
HST03X	0.216	0.0536	−0.00401	−0.00654	−0.241	−0.00579	0.0924	0.0456	−0.0809	−0.0573
HST79X	0.142	0.155	−0.165	0.158	0.177	−0.00807	−0.116	−0.0667	0.0578	0.069
MODLYR	0.0461	−0.277	−0.155	0.0782	0.0175	0.021	−0.00773	−0.0224	−0.00894	0.144
OREVTR	0.216	−0.05	0.138	−0.012	−0.088	−0.00707	0.0534	0.0194	−0.0401	0.14
ORVTB0	0.161	−0.0807	0.188	0.0113	−0.21	−0.00938	0.104	0.0425	−0.113	0.0978
REHSAT	0.152	−0.0309	0.0739	−0.0214	−0.189	0.011	0.0666	0.00493	0.119	0.28
RVOLDT	0.212	0.00907	−0.00634	−0.16	0.0124	0.0202	0.00774	−0.0446	0.299	0.135
RVTRDS	0.238	0.0395	0.0107	−0.018	−0.0835	−3.5E-4	0.0226	0.0133	−0.0395	0.0882
T2924X	0.159	0.177	−0.145	0.261	0.0633	−0.0062	−0.0681	−0.0205	−0.0711	0.0236
T3924X	0.143	0.185	−0.16	0.269	0.0996	−0.00442	−0.0854	−0.0335	−0.0663	0.0406
T4924X	0.131	0.19	−0.171	0.269	0.117	−0.00613	−0.0943	−0.0394	−0.0628	0.0507
TIME29	0.0192	0.0166	−0.184	−0.351	0.178	−0.0108	0.0379	0.0514	−0.304	0.0122
TIME39	0.0289	0.0121	−0.182	−0.363	0.176	−0.0154	0.0478	0.0551	−0.31	−0.0064
TIME49	0.0379	0.00823	−0.175	−0.355	0.167	−0.0126	0.0488	0.0562	−0.291	−0.0153
TROP24	0.147	0.168	−0.127	0.248	0.00868	−0.071	−0.0248	0.00595	−0.165	0.00173
NEWUSE = U	0.00823	−0.017	−0.00211	0.00907	−0.0212	0.698	−0.0518	−0.0063	−0.0472	−0.0471
NEWUSE = N	−0.00823	0.017	0.00211	−0.00907	0.0212	−0.698	0.0518	0.0063	0.0472	0.0471
VDDASAV = DDA	0.0327	−0.0256	0.00231	−0.0143	−0.063	0.00334	−0.257	0.434	0.081	0.0426
VDDASAV = NONE	−0.0342	0.0437	0.0145	0.0138	0.0267	0.0221	0.345	−0.641	−0.0876	0.0244
VDDASAV = SAV	−0.0132	−0.0129	−0.0073	−0.0036	0.0498	−0.0228	−0.164	0.301	0.0146	−0.076
VDDASAV = BOTH	0.0317	−0.0305	−0.0229	−8.49E-4	−0.0139	−0.0212	−0.0872	0.233	0.0299	−0.0233

Figure 5.9   Principal component analysis of the credit scoring data set

Rotated factors:

Principal components	PC1(rot)	PC2(rot)	PC3(rot)	PC4(rot)	PC5(rot)	PC6(rot)	PC7(rot)	PC8(rot)	PC9(rot)	PC10(rot)
AGEAVG	−0.0287	−0.00342	−0.0402	−0.0506	0.0295	−0.0077	−0.00232	−0.0024	0.429	0.0397
AGEOTD	0.00257	0.00673	−0.0292	−0.0122	0.00734	−2.97E-4	0.0139	−0.00741	0.445	−0.0441
AUUTIL	0.00174	0.0116	−0.0182	0.0551	−0.136	0.00496	0.0178	0.0245	−0.0634	−0.321
BAD	0.00179	−0.0028	7.69E-4	−0.00213	0.00313	−2.35E-4	0.683	−0.0032	0.00485	5E-4
BKTIME	−0.00777	0.00519	0.0385	−0.0133	0.08	0.00697	−0.0306	0.0319	0.151	−0.19
BRHS2X	0.00781	6.46E-4	0.408	−0.0136	0.0192	0.0031	0.0013	0.00116	−0.00865	−0.0135
BRHS3X	0.0114	5.55E-4	0.41	−0.0127	0.0256	0.00358	3.14E-4	0.00149	−0.00805	−0.0125
BRHS4X	0.011	0.00139	0.412	−0.0123	0.0292	0.00443	0.0014	5.1E-4	−0.00667	−0.0114
BRHS5X	0.013	0.00105	0.409	−0.00937	0.0302	0.0037	4.21E-4	−0.00164	−0.00166	−0.012
BROLDT	−0.00359	−0.00109	0.203	0.00263	0.0244	−0.00303	−0.00185	0.00183	0.263	0.0021
BROPEN	0.00185	0.00167	0.39	−0.024	−0.0359	0.0027	6.93E-4	0.00112	−0.0279	−0.0127
BRTRDS	0.126	0.00769	0.119	−0.0913	−0.155	0.0102	−0.0139	0.0085	0.0495	0.014
BSRETL	−0.00505	−0.393	0.0015	−0.00482	−0.0139	0.0028	0.00738	−1.75E-4	−0.0191	−0.0752
BSWHOL	−0.00513	−0.394	−3.22E-4	−0.00339	−0.0142	0.00489	0.00983	0.00459	−0.0175	−0.0681
CBTYPE	−0.212	0.00182	0.103	0.217	0.108	−0.0101	0.00706	−0.0142	8.03E-4	0.0492
CFTRDS	0.136	0.0205	−0.0568	−0.023	−0.179	0.00474	−0.0239	−0.0069	0.0428	−0.189
CONTPR	0.00496	−0.407	0.00185	−0.00558	0.0117	−0.00998	0.0195	0.0123	0.00436	0.00705
CURSAT	0.066	2.4E-4	−0.0578	−0.0986	−0.375	−0.00228	−0.00582	−0.00137	−0.0184	−0.0141
DWNPMT	−0.0153	−0.262	0.0364	0.0227	0.0469	−0.0353	−0.0136	−0.044	−0.0297	−0.346
GOOD	−0.00179	0.0028	−7.69E-4	0.00213	−0.00313	2.35E-4	−0.683	0.0032	−0.00485	−5E-4
HSATRT	−0.0819	0.00315	−0.0835	−0.0703	−0.384	0.0163	−0.0244	−8.73E-4	−0.0214	−0.06
INQ012	0.0856	−0.0429	−0.00738	−3.07E-4	−0.109	−0.0454	0.065	6.92E-5	−0.0742	−0.0879
MILEAG	3.75E-4	0.216	0.00823	−0.00901	−0.00287	−0.0193	0.00718	−0.0257	−0.0521	−0.197
MNGPAY	−0.0167	−0.0278	−0.00716	−0.0218	−0.0365	0.00522	0.00253	0.0231	−0.0148	−0.107
NTTRIN	0.042	−0.0484	−0.0325	−0.0485	−0.0339	0.00075	−0.0274	0.0229	0.0401	0.427
PUBREC	0.151	−0.00327	−0.0359	−0.118	0.0315	0.0383	−0.00469	0.0312	0.112	−0.2
TERM	0.00681	−0.356	−0.0156	−7.43E-4	0.00297	0.00966	0.0101	−0.0135	0.0199	0.127
TOTBAL	0.0307	−0.00455	0.0103	0.0828	−0.105	0.0143	−0.00425	0.0402	0.0954	−0.234
TRADES	0.25	0.00298	0.0332	0.0767	−0.0864	−0.00779	−0.0105	−0.0021	0.108	−0.0106
VJOBMOS	−0.0648	−0.0126	0.00711	0.00731	−0.0352	4.23E-4	−0.0204	0.0112	0.094	−0.362
VRESMOS	−0.0697	−0.00575	0.00168	0.0469	0.0106	0.0217	−0.0344	−0.0549	0.0799	−0.253
AGEOTD	0.00257	0.00673	−0.0292	−0.0122	0.00734	−2.97E-4	0.0139	−0.00741	0.445	−0.0441
BKRETL	−0.00276	−0.407	7.75E-4	−0.00244	−0.00778	−0.00284	0.00477	0.00537	−0.00313	−0.0354
BRBAL1	−0.1	−0.00141	0.21	0.0302	−0.193	−0.00167	2.65E-4	−0.00769	−0.108	0.01
CSORAT	−0.0553	−0.01	−0.03	−0.0112	−0.342	−0.00272	0.00468	−0.0103	0.00399	0.0293
HST03X	0.055	0.0154	−0.025	−0.00197	−0.341	0.00489	−0.00807	0.00681	0.0158	−0.0881
HST79X	0.346	−0.00538	0.0305	0.0192	0.125	−0.0119	−0.0101	−0.00651	0.122	0.0442
MODLYR	0.0165	−0.329	−0.0111	8.4E-4	0.00813	0.0203	−0.0412	−0.00307	0.0137	0.14
OREVTR	−0.0159	−0.0127	0.143	0.0374	−0.239	−0.0128	0.00543	−0.00142	0.0557	0.134
ORVTB0	−0.0869	−0.00959	0.101	0.0655	−0.332	−0.0077	0.00387	−0.0105	−0.065	0.102
REHSAT	−0.0538	−0.017	−0.0258	0.138	−0.231	−0.0137	0.0181	−0.00304	0.171	0.238
RVOLDT	−0.0353	−4.02E-4	0.0412	0.0575	−0.0418	−0.00879	0.0141	0.00387	0.0408	0.0741
RVTRDS	0.0908	0.00321	0.0663	−0.0186	−0.22	−0.00127	−0.0138	0.00814	0.0977	0.0653
T2924X	0.396	−0.00125	0.00963	0.0439	−0.0128	0.0017	0.00626	−0.00152	−0.0305	0.00668
T3924X	0.416	−0.00119	0.00913	0.0387	0.0308	0.0021	0.00421	−0.00723	−0.0268	0.0256
T4924X	0.424	−3.85E-4	0.00332	0.0324	0.0549	−5.37E-4	0.00273	−0.00938	−0.023	0.036
TIME29	−0.0152	0.00121	0.00555	−0.534	0.0071	−0.00234	8.7E-4	−0.00616	4.22E-4	0.0306
TIME39	−0.0228	−0.00111	0.0125	−0.546	−0.00504	−0.00578	0.0035	−0.00669	0.00108	0.0116
TIME49	−0.0257	−0.00405	0.0161	−0.525	−0.0108	−0.00331	0.00502	−0.00298	0.0115	4.07E4
TROP24	0.368	0.00377	−0.0013	5.25E-4	−0.0881	−0.00333	0.0138	−0.0074	−0.113	−0.00694
NEWUSE = U	−4.82E-4	−4.79E-4	0.00156	0.00226	−4.48E-4	0.704	2.66E-4	−5.76E-4	−0.00186	0.0016
NEWUSE = N	4.82E-4	4.79E-4	−0.00156	−0.00226	4.48E-4	−0.704	−2.66E-4	5.76E-4	0.00186	−0.0016
VDDASAV = DDA	−0.00954	0.0101	−0.0177	0.0305	−0.0248	0.0131	−0.0231	0.513	0.0221	0.0519
VDDASAV = NONE	0.00192	0.00351	−0.00337	8.8E-4	−0.00492	0.0038	−0.0012	−0.737	0.0061	0.00534
VDDASAV = SAV	0.00711	0.00708	0.029	−0.0346	0.0588	−0.0101	0.0229	0.342	−0.0406	−0.0568
VDDASAV = BOTH	0.00245	−0.0299	−2.75E-4	−0.00766	−0.0207	−0.0153	0.0121	0.253	0.0018	−0.0228

Figure 5.9   (*Continued*)

The following rules were used to eliminate outliers: *CFTRDS* greater than 30, *mileag* greater than 300,000, *MNGPAY* greater than 100,000, *VRESMOS* greater than 1000, and *TROP24* greater than 50. An alternative approach is to cap the values, for example, assigning all individuals with pay greater than $100,000 to a $100,000 cap.

5.5.3 Analysis

Segmentation   To facilitate the analysis, a randomly generated subset of 5000 observations was generated.

Principal Component Analysis   To better understand the factors associated with the variation in the data set, a principal component analysis was performed including a factor rotation as shown in Fig. 5.9.

From this PCA analysis, the groups shown in Table 5.12 were identified, and given a name, in order to understand the variation in the data set. One variable from each of the groups was selected: *T4924X*, *BSRETL*, *BRHS4X*, *TIME39*, and *ORVTB0*.

Creating Discrete Variables   As discussed earlier, many of the variables, including most of the five selected, contain negative values which have specific meanings. To remove all observations with the negative values would be problematic for a number of reasons. It would reduce the number to be used for training the models. In addition, eliminating cases with the negative values would reduce the utility of the model, since the deployed model could only be used with cases that have positive values. Instead of removing the values, the variables are binned based on ranges of values. All negative values are placed in one of the bins. The divisions are chosen to generate groups such that (1) the groups divide observations where *good* = 1 and *good* = 0, and (2) will contain at least 2% of the data set. Table 5.13 illustrates the variables selected, and the grouping of the variables.

Prioritized Variables   To further understand the relationship between the potential independent variables and the response variable, a number of statistical tests were

TABLE 5.12 Description of the First Five Principal Components

Principal component	Factor names	Variables
PC-1	"Trade activity"	*TRADES, T2924X, T3924X, T4924X, TROP24, HST79X*
PC-2	"Auto price"	*BSRETL, BSWHOL, CONTR, DWNPAY, TERM, BKRETL, MODLYR*
PC-3	"Revolving trade activity"	*BRHS2X, BRHS3X, BRHS4X, BRHS5X, BROPEN, BRBALI*
PC-4	"Days past due rating"	*CBTYPE, TIME29, TIME39, TIME49*
PC-5	"Satisfactory trade"	*CURSAT, HSATRT, CSORAT, HST03X, OREVTR, ORVTB0, REHSAT*

TABLE 5.13 Binning of the Selected Variables

Variables	Binning
T4924X	<0; 0; $1-2$; $3-9$; $9+$
BSRETL	$0-4000$; $4000-6000$; $6000-15,000$; $15,000+$
BRHS4X	<0; 0; $1+$
TIME39	<0; 0; $1-9$; $10+$
ORVTB0	<0; 0; $1-2$; $3+$

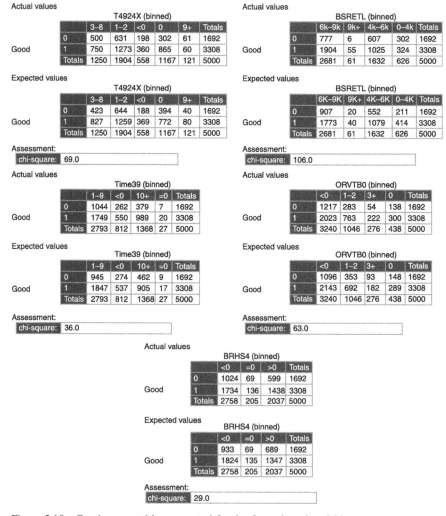

Figure 5.10 Contingency tables generated for the five selected variables

performed and summarized in Fig. 5.10. For each variable, the chi-square assessment is above the critical value, indicating the existence of a relationship.

Dummy Variables   A dummy variable was generated from each of the binned values of selected variables, as described in Table 5.14. For the dummy variables derived from *T4924X*, *BRHS4X*, *TIME39*, and *ORVTB0*, the variables containing values less than zero are not considered to avoid colinearity issues. Similarly, the new dummy variable *BSRETL (15000+)* is not used.

Building Logistic Regression Models   To further facilitate the analysis, a new random subset of 1000 observations was created. Since the response variable is binary, the logistic regression model building approach appears appropriate. Combinations of independent variables were used to build multiple models. A 10% cross-validation was used to ensure that no observation's prediction was calculated from a model in which that observation was used to train the model. Where the predicted probability is greater than 0.65, the prediction is assigned 1, otherwise it is assigned 0. Some results are shown in Table 5.15.

The model presented in Fig. 5.11 was selected, because of its overall accuracy, sensitivity, and specificity values.

In this model, the false discovery rate is considered important. In general, 32% of all loan applicants will default on their loans. Using this model, the number has been reduced to 26.5%, or the 140 false positives divided by the 529 predicted

TABLE 5.14   Dummy Variables Generated

Original variables	New dummy variables
T4924X	*T4924X (binned) < 0*
	T4924X (binned) = 0
	T4924X (binned) = 1–2
	T4924X (binned) = 3–9
	T4924X (binned) = 9+
BSRETL	*BSRETL (binned) = 0–4k*
	BSRETL (binned) = 4k–6k
	BSRETL (binned) = 6k–15k
	BSRETL (binned) = 15k+
BRHS4X	*BRHS4X (binned) < 0*
	BRHS4X (binned) = 0
	BRHS4X (binned) = 1+
TIME39	*TIME39 (binned) < 0*
	TIME39 (binned) = 0
	TIME39 (binned) = 1–9
	TIME39 (binned) = 10+
ORVTB0	*ORVTB0 (binned) < 0*
	ORVTB0 (binned) = 0
	ORVTB0 (binned) = 1–3
	ORVTB0 (binned) = 3+

TABLE 5.15 Building Multiple Logistic Regression Models from Combinations of Independent Variables

ACC	SENS	SPEC	V1	V2	V3	V4	V5	V6
0.596	0.597	0.596	T4924X (binned) = 9+	BSRETL (binned) = 4k–6k	BSRETL (binned) = 0–4k			
0.596	0.597	0.596	T4924X (binned) = 1–2	T4924X (binned) = 9+	BSRETL (binned) = 4k–6k	BSRETL (binned) = 0–4k		
0.596	0.597	0.596	T4924X (binned) = 9+	BSRETL (binned) = 6k–9k	BSRETL (binned) = 4k–6k	BSRETL (binned) = 0–4k		
0.596	0.597	0.596	T4924X (binned) = 1–2	T4924X (binned) = 9+	BSRETL (binned) = 6k–9k	BSRETL (binned) = 4k–6k	BSRETL (binned) = 0–4k	
0.598	0.582	0.606	BSRETL (binned) = 4k–6k	BSRETL (binned) = 0–4k				
0.598	0.582	0.606	T4924X (binned) = 1–2	BSRETL (binned) = 4k–6k	BSRETL (binned) = 0–4k			
0.598	0.582	0.606	BSRETL (binned) = 6k–9k	BSRETL (binned) = 4k–6k	BSRETL (binned) = 0–4k			
0.598	0.582	0.606	T4924X (binned) = 1–2	BSRETL (binned) = 6k–9k	BSRETL (binned) = 4k–6k	BSRETL (binned) = 0–4k		
0.598	0.582	0.606	BSRETL (binned) = 4k–6k	BSRETL (binned) = 0–4k	ORVTB0 (binned) = 0			
0.574	0.545	0.59	BRHS4 (binned) = >0	BRHS4 (binned) = 0	TIME39 (binned) = 1–9	TIME39 (binned) = 0	ORVTB0 (binned) = 1–2	ORVTB0 (binned) = 3+
0.579	0.507	0.617	BRHS4 (binned) = >0	TIME39 (binned) = 1–9	TIME39 (binned) = 10+	ORVTB0 (binned) = 1–2	ORVTB0 (binned) = 3+	ORVTB0 (binned) = 0
0.567	0.545	0.579	TIME39 (binned) = 1–9	TIME39 (binned) = 0	ORVTB0 (binned) = 1–2	ORVTB0 (binned) = 3+		
0.559	0.573	0.551	BRHS4 (binned) = 0	TIME39 (binned) = 1–9	TIME39 (binned) = 10+	TIME39 (binned) = 0	ORVTB0 (binned) = 1–2	
...	...	...	...	...	...	...	...	...

Logistic regression model summary	
Descriptors	T4924X (binned) = 9+, BSRETL (binned) = 4–6k, BSRETL (binned) = 0–4k
Response	good

Cross validated results	
Accuracy	0.596
Error	0.404
Sensitivity	0.596
Specificity	0.597
False positive rate	0.403
Positive predictive value	0.735
Negative predictive value	0.439
False discovery rate	0.265

Actual (good)

	1	0	Totals
Predicted (good) 1	389	140	529
0	264	207	471
Totals	653	347	1000

Coefficients	Coefficients
Intercept	1.02
T4924x (binned) = 9+	−0.736
BSRETL (binned) = 4–6k	−0.687
BSRETL (binned) = 0–4k	−0.986

Figure 5.11   Summary of the logistic regression model selected

positives. The benefits of using this model are the amount of money saved per percentage point improvement in the false discovery rate.

5.5.4   Deployment

A number of steps were taken in preparing the training set data. These steps should be repeated on any new data to which this model is to be applied.

5.6   DATA MINING NONTABULAR DATA

5.6.1   Overview

In many situations, the data to be mined is not in a tabular format, amenable to mining using the methods described in this book. In order to look for trends and patterns as well as to build predictive models from this data, the information must be pre-processed in order to generate the required formatted data. The following section describes two problem areas where the original data is not in a tabular format: mining chemical data and mining text data. In both cases, the original data is first transformed into a table of data, and then techniques detailed earlier in the book can be used to effectively mine the data.

5.6.2   Data Mining Chemical Data

Chemical data presents a number of unique challenges when attempting to mine data sets incorporating this information. An example of a chemical is shown in Fig. 5.12. The picture represents the atoms and bonds of this particular chemical. The chemical

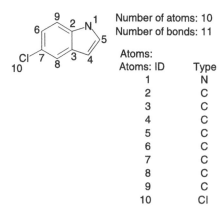

Figure 5.12   Example of a chemical

can be considered a graph, with the vertices of the graph comprising the atoms and the connections between the vertices representing the bonds of the chemicals. Vertices with no symbols are carbons (symbol C), and vertices with letters represent other atoms, for example, the symbol N represents a nitrogen atom, and the symbol Cl represents a chlorine atom. The type of a bond is also shown on the drawing. In this example, single bonds are represented as a single line between the atoms, and double bonds are represented as two lines between the atoms.

The connections between the atoms and bonds of a single chemical are usually represented on a computer as a connection table. Figure 5.13 illustrates the atom and bond information that are incorporated into a connection table. One of the more popular formats for representing chemical structures is the MDL Molfile (Dalby et al., 1992), which includes a standardized representation of a chemical's connection table, as shown in Fig. 5.14. The file includes the number of atoms and bonds in the chemical, a listing of the atoms (along with associated information), and a listing of the bonds (along with associated information) and their connections to the atoms.

Number of atoms: 10
Number of bonds: 11

Atoms:

Atoms: ID	Type
1	N
2	C
3	C
4	C
5	C
6	C
7	C
8	C
9	C
10	Cl

Bonds:

From atom	To atom	Type
5	1	Single (1)
2	3	Single (1)
6	7	Double (2)
1	2	Single (1)
7	8	Single (1)
8	3	Double (2)
3	4	Single (1)
2	9	Double (2)
9	6	Single (1)
4	5	Double (2)
7	10	Single (1)

Figure 5.13   Connection table representing the chemical

Chemical Structure-1

```
10 11  0  0  0  0  0  0  0  0  0.999  V2000
    3.8226    -0.4964    0.0000  N   0 0 0 0 0 0 0 0 0 0 0 0
    3.0380    -0.7541    0.0000  C   0 0 0 0 0 0 0 0 0 0 0 0
    3.0418    -1.5806    0.0000  C   0 0 0 0 0 0 0 0 0 0 0 0
    3.8297    -1.8311    0.0000  C   0 0 0 0 0 0 0 0 0 0 0 0
    4.3101    -1.1585    0.0000  C   0 0 0 0 0 0 0 0 0 0 0 0
    1.6194    -0.7500    0.0000  C   0 0 0 0 0 0 0 0 0 0 0 0
    1.6141    -1.5732    0.0000  C   0 0 0 0 0 0 0 0 0 0 0 0
    2.3232    -1.9875    0.0000  C   0 0 0 0 0 0 0 0 0 0 0 0
    2.3297    -0.3428    0.0000  C   0 0 0 0 0 0 0 0 0 0 0 0
    0.8958    -1.9792    0.0000  Cl  0 0 0 0 0 0 0 0 0 0 0 0
  5  1  1  0  0  0  0
  2  3  1  0  0  0  0
  6  7  2  0  0  0  0
  1  2  1  0  0  0  0
  7  8  1  0  0  0  0
  8  3  2  0  0  0  0
  3  4  1  0  0  0  0
  2  9  2  0  0  0  0
  9  6  1  0  0  0  0
  4  5  2  0  0  0  0
  7 10  1  0  0  0  0
M   END
```

Figure 5.14   An MDL Molfile computer representation of a chemical

In order to mine data sets containing chemicals, the computer representation of the chemical structure is usually cleaned to take into account possible alternative ways of drawing the same chemical. For example, the two chemicals in Fig. 5.15 represent the same chemical, even though the bonds are in a different order.

Once the chemical representation has been appropriately cleaned and normalized, the information can be further processed to generate numeric independent variables that describe either the whole chemical or parts of the chemical. Some of the most popular independent variables to describe chemical structures are molecular weight, number of rotatable bonds, log$P$ (a prediction that relates to how well a chemical is transported throughout the body), and substructural fragments. Rotatable bonds and substructural fragments will be used to illustrate how these independent variables are generated.

- *Number of rotatable bonds*: This is a numeric property of a chemical that indicates a chemical's flexibility. It is a single count of the number of bonds about which the chemical can freely rotate. Generally, the more bonds the chemical can rotate around, the more flexible a chemical is. Rotatable bonds are usually defined as bonds that are single, not part of a ring, and not on the outer extremities of the chemical (see Fig. 5.16 for examples).

Figure 5.15   Different ways of drawing the same chemical structure

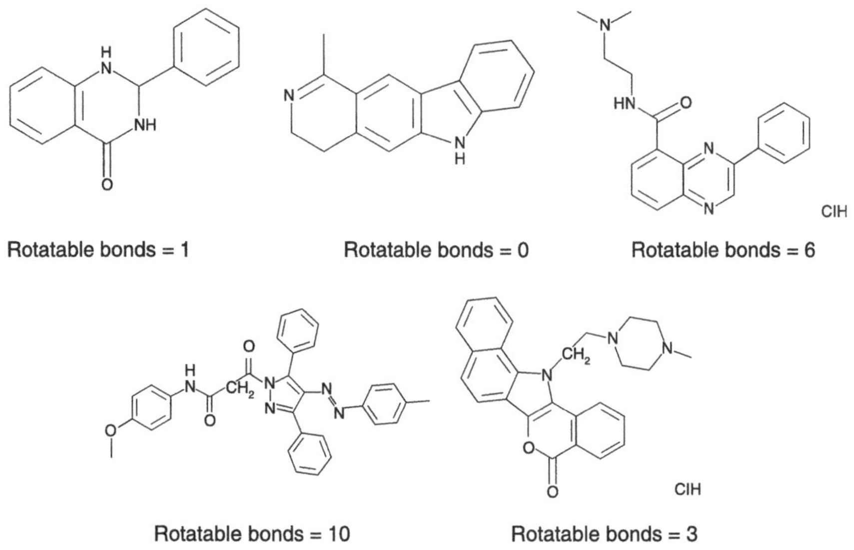

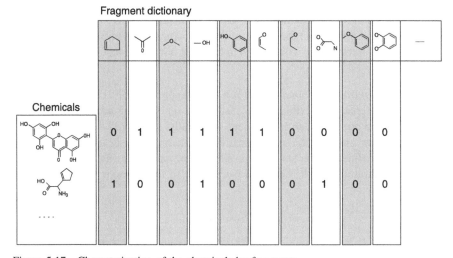

Rotatable bonds = 1   Rotatable bonds = 0   Rotatable bonds = 6

Rotatable bonds = 10   Rotatable bonds = 3

Figure 5.16   Calculation of number of rotatable bonds

- *Substructural fragments*: A chemical can be characterized through an indication of the presence or absence of small chemical pieces. An example is shown in Fig. 5.17. Here a software program systematically looks for the presence or absence of the small molecular fragments shown in the column headings. A 1 in the main body indicates that particular fragment is present somewhere in the chemical and a 0 indicates that the fragment is not present.

Figure 5.17   Characterization of the chemicals by fragments

The generation of independent variables is an important part of any exercise to mine chemical data. The resulting data tables allow for the application of data mining methods, such as clustering and predictive analytics, to problems involving chemicals.

Two examples will be used to illustrate how a set of chemicals along with associated information can be processed:

- *Identifying diverse chemicals*: In many situations, a pharmaceutical or chemical company needs to identify a subset of chemicals from a large collection that are diverse, meaning that they are not structurally similar to others. One approach is to initially generate a table of independent variables similar to Fig. 5.17 and then use clustering to group the chemicals into sets of structurally similar chemicals. Once clustered, a representative from each cluster can be selected. To illustrate the process, the set of seven chemicals in Fig. 5.18 is used. These chemicals are clustered using agglomerative hierarchical clustering, as discussed earlier in this book. Figures 5.19 and 5.20 show the clustering of these seven chemicals, with a cutoff set to form three groups. At this point a representative chemical may be selected from each cluster to create a diverse set. In this example, selecting chemicals B, G, and D would represent a subset of diverse chemicals. The seven chemicals in this example are used to illustrate the process; however, in practice this approach is usually applied to thousands of chemicals.

- *Understanding drug discovery data*: Throughout the early stages of drug discovery, chemicals are tested, often in test tubes, to attempt to identify

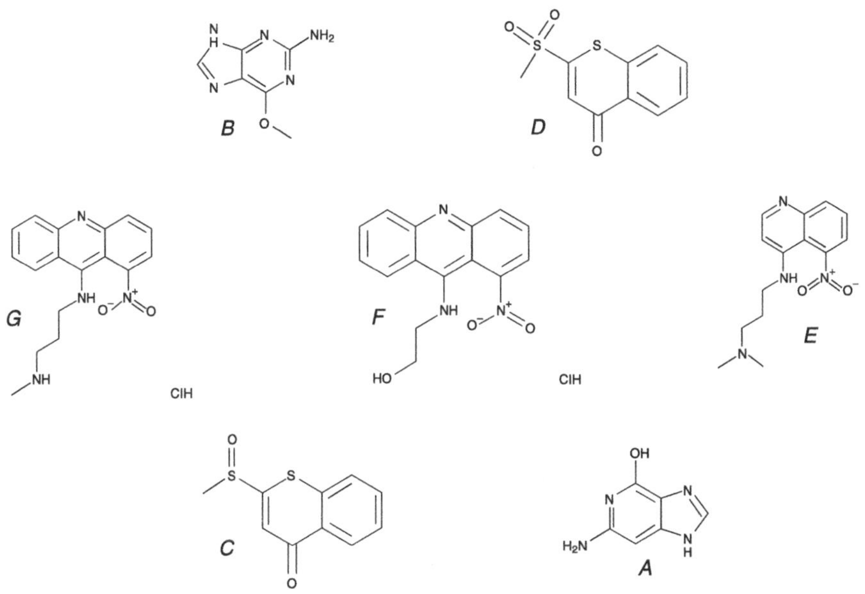

Figure 5.18   Seven chemicals

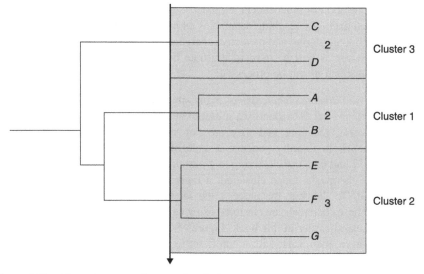

Figure 5.19 Clustering a set of seven chemicals

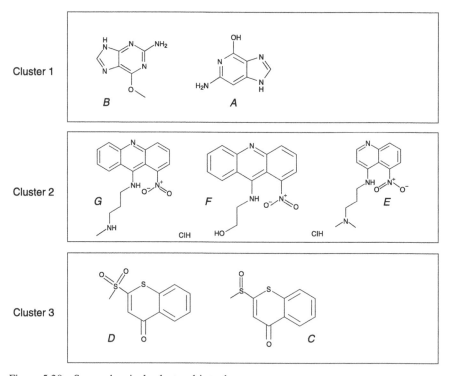

Figure 5.20 Seven chemicals clustered into three groups

TABLE 5.16   Seven Chemicals Tested for Cancer

Structure ID	Data
A	4.44
B	3.05
C	5.21
D	5.04
E	5.23
F	9.33
G	8.71

potential drug candidates. These experiments result in data concerning the potential biological effects of these individual chemicals, such as their ability to treat a particular disease. Using independent variables, such as the number of rotatable bonds or substructural fragments, this data can be mined using the biological data as a response variable. To illustrate, the seven molecules in Fig. 5.18 have been tested for a specific cancer, resulting in the data table in Table 5.16. The higher the value, the better the chemical is expected to be at treating cancer. A set of independent variables is generated in Fig. 5.21. This information is data mined using these dummy variables as independent variables and the biological data as the response. Predictive models can now be built that could be used to predict the propensity of other chemicals for treating cancer. Descriptive data mining, such as clustering and decision trees, can be built to help understand what properties of the chemical appear to be responsible for treating cancer. For example, the decision tree in Fig. 5.22 shows that the presence of the three ring fragment is particularly important. Again, in practice these approaches are applied to large databases of chemicals and associated biological data.

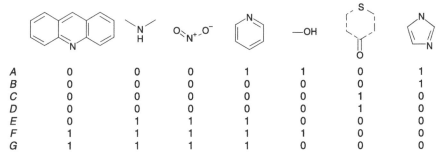

A	0	0	0	1	1	0	1
B	0	0	0	0	0	0	1
C	0	0	0	0	0	1	0
D	0	0	0	0	0	1	0
E	0	1	1	1	0	0	0
F	1	1	1	1	1	0	0
G	1	1	1	1	0	0	0

Figure 5.21   Independent variables generated for the seven molecules

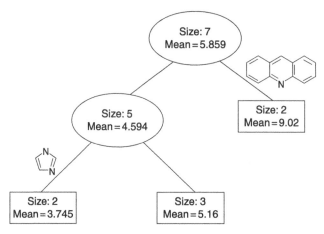

Figure 5.22 Decision tree for a set of chemical structures

5.6.3 Data Mining Text

Text mining takes ideas from the data mining of highly structured, mostly numeric databases or data warehouses into the world of semistructured or unstructured electronic documents such as email, memoranda, articles, or HTML and XML files. Humans have an ability to read and interpret the text documents that has been difficult to automate with machine intelligence. Computers do not easily overcome textual barriers such as slang, synonyms, spelling variations, and contextual meaning. However, advances in information retrieval, natural language processing, and corpus-based computational linguistics have enabled text mining technologies to extract information and track topics; to summarize, categorize, and cluster documents; and to conceptually link similar documents even when the words in the documents may be different.

There is disagreement on definition and what qualifies as text mining. The discipline of information retrieval focuses on finding and retrieving what is already known. Some argue that technologies like information extraction that identify and tag, for example, people, places, events, and times from a passage in a document find what is already in the text and are therefore just advanced information retrieval. They argue that, to qualify as text mining, the technology must learn something new about the world that is completely outside the document collection being analyzed (*knowledge creation*). Others relax that requirement and argue that, although the technology may not find something completely novel, it must at least find something unknown such as new patterns or trends (*knowledge discovery*). But all agree that text mining is a close relative of data mining and you will find many similarities throughout: from the way the text is processed to the technologies used to process it. To build on what has been covered so far and to provide an appreciation for processing text, the following will focus on the clustering technology. The general architecture of a text mining system is initially discussed.

The general architecture shown in Fig. 5.23 should feel familiar. In place of a data set you find a *document collection*; in place of an observation you find a

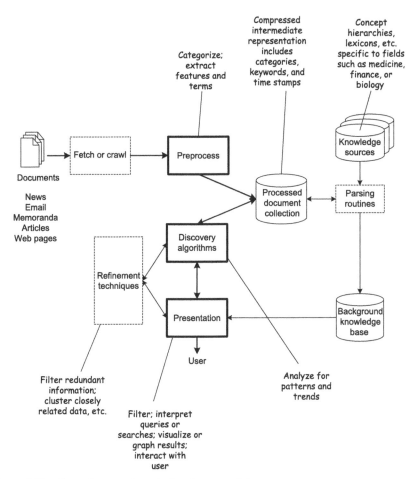

Figure 5.23   General processing pipeline for text mining

document; in place of attributes or multiple variables you find *document features*. At a high level, the text processing pipeline takes as input text documents and produces as output patterns, trends, or connections that can be visualized or browsed. Text processing is done in three major stages: The preprocessing stage creates an intermediate representation of the documents that can be used by core mining operations. These operations, along with refinement techniques, interact with the presentation layer to allow the user to visualize or explore the results.

Clustering, as discussed in Chapter 3, groups things that are similar. A major reason for its use in certain text mining applications is that documents relevant to a particular search tend to be more similar to each other than nonrelevant ones. Clustering differs from categorization in that the grouping is done on-the-fly rather than assigning documents to predefined topics or categories, as is done for example when filtering spam. Some of the ways clustering can be applied are to improve search *recall*, search *precision*, or to organize the results of a search into a concept hierarchy.

Recall is a technical term from information retrieval. It is the proportion of all relevant documents retrieved out of all relevant documents available. For example, if the search query used the term "car," documents containing terms like "auto" or "automobile" that were not retrieved would result in a lower recall. If the search returned 15 relevant documents but should have returned 20, the recall was $15/20$ or 0.75. Clustering helps find these documents by using other words in the text, since documents about "cars" or "autos" tend to have similar words overall.

Precision is also a technical term from information retrieval. It is the proportion of the number of relevant documents out of all documents retrieved. For example, a query with the term "Saturn" intended to retrieve information about cars would return documents about the General Motors car and the planet. If the search returned 20 documents but only 10 were relevant to cars (the intent of the search), the precision is $10/20$ or 0.50. Clustering helps improve precision by grouping the documents into smaller sets and returning only the most relevant groups.

Clustering can also be used to organize the results of a search, creating on-the-fly a concept hierarchy or dynamic table of contents of the documents retrieved. You can see this in action by using the clusty search engine (clusty.com). Type in the query "chemical" and along with the results you may see something similar to the following concept hierarchy, which provide links to the documents in the cluster labeled by the concept:

+ Engineering (40)

+ Chemical Industry (14)

+ My Chemical Romance (13)

+ Safety (13)

+ Chemical suppliers (10)

+ Definitions, Dictionary (7)

+ Chemical Company (10)

+ Printing (9)

+ UK, Specialty Chemicals (8)

+ China (6)

A basic clustering algorithm for documents requires the following:

- a representation of the document;
- a distance metric;
- a clustering method.

Document Representation   In the *vector space model*, documents must first be converted into vectors in the *feature space*. The most common features are the unique terms (bag of words) found across all documents in the document collection. Each document's vector will have a component for each feature in the space. The component will be some positive number if the document has that feature or 0 otherwise. The most popular weighting scheme multiplies the frequency of the term in the document with the inverse of the frequency of the term in the collection ($tf * idf$). This metric is used because it is thought that a term that appears much more frequently in the document than in the collection is a better discriminator. So a document d_i might be represented as:

$$d_i = (tf_1 * idf_1, tf_2 * idf_2, \ldots, tf_m * idf_m)$$

where m is the number of features in the feature space. All vectors are normalized to the unit vector:

$$d'_i = d_i / |d_i|$$

Distance Metric   The most commonly used distance metric used to compare documents is the cosine measure discussed in Chapter 3. Because a document collection can have a very large number of unique terms, steps are taken in the preprocessing to reduce the number of terms that will be used as features in the vector model. *Filtering* removes punctuation and special characters not considered important. *Tokenization*, the splitting of the text into chunks, might use phrases, such as noun phrases, rather than just individual words. *Stemming* reduces words to their roots so that, for example, "state," "states," and "stated" all become "state." *Stopwords* like "the," which carry no content, are eliminated. *Pruning* removes all terms with very low frequency since they would likely produce clusters too small to be of use.

Clustering Method   For the clustering method, one of the various clustering algorithms discussed in Chapter 3 may be used. Historically, partitioned-based methods such as k-means (using the cosine distance metric) or hierarchical methods such as agglomerative clustering have been popular (Fox, 2007). Hierarchical clustering is reported not to scale well and k-means, although easy to implement are reported to have other issues: random initialization, convergence to suboptimal local minima, and the complexity of the algorithm, which is $O(n \cdot k \cdot l)$ where n is the number of documents in the collection, k is the number of clusters, and l is the number of iterations (Fox, 2007).

5.7   FURTHER READING

Davenport and Harris (2007) provide an overview of the use of analytics to gain a competitive advantage across a range of industries. There are a number of publications that discuss generally the application of data mining, including Klösgen and Żytkow (2002), Maimon and Rokach (2005), Kantardzic and Zurada (2005), Alhaji et al. (2007), Li et al. (2005, 2006), Sumathi and Sivanandam (2006), and Perner (2006). In addition, many software vendors discuss the application of data mining on their websites, and a list of vendors may be found at http://www. kdnuggets.com/companies/products.html. There are additional resources covering data mining applications in the areas of sales and marketing (Berry and Lindoff,

2004; Rud, 2001), finance (McNeils, 2004; Shadbolt and Taylor, 2002), manufacturing (Braha, 2001), security applications (Thuraisingham, 2003; Mean, 2003; and McCue, 2007), science and engineering (Grossman et al., 2001), and healthcare (Fielding, 2007; Pardalos et al., 2007; Chen et al., 2005; Berner, 2006; and Shortliffe and Cimino, 2006). General information on the fields of bioinformatics are covered in Xiong (2006), Baxevanis and Francis Ouellette (2005), and Jones and Pevzner (2004), and for chemoinformatics Gasteiger and Engel (2003) and Leach and Gillet (2003). Additional case studies of data mining applications are found in Guidici (2003). Fan et al. (2006) provides an introduction to text data mining, Hearst (1999) defines text data mining concepts and terminology, and Andrews and Fox (2007) provide a recent overview of research and advanced development in document clustering.

MATRICES

A.1 OVERVIEW OF MATRICES

A matrix represents a table of quantities. It is defined by the number of rows and columns, along with the individual values. For example, the following matrix, X, has three rows and four columns:

$$X = \begin{bmatrix} 3 & 5 & 2 & 4 \\ 7 & 6 & 8 & 4 \\ 5 & 1 & 3 & 8 \end{bmatrix}$$

The following appendix outlines a number of common matrix operations: addition, multiplication, transpose, and inversion.

A.2 MATRIX ADDITION

It is possible to add two or more matrices together, where they have the same number of rows and columns. This is achieved by adding together the individual elements from each matrix. Matrices X and Y are added together in this example.

$$X = \begin{bmatrix} 3 & 5 & 2 & 4 \\ 7 & 6 & 8 & 4 \\ 5 & 1 & 3 & 8 \end{bmatrix}$$

$$Y = \begin{bmatrix} 2 & 6 & 3 & 7 \\ 9 & 3 & 8 & 1 \\ 4 & 6 & 9 & 7 \end{bmatrix}$$

Making Sense of Data II. By Glenn J. Myatt and Wayne P. Johnson
Copyright © 2009 John Wiley & Sons, Inc.

$$X + Y = \begin{bmatrix} 3 & 5 & 2 & 4 \\ 7 & 6 & 8 & 4 \\ 5 & 1 & 3 & 8 \end{bmatrix} + \begin{bmatrix} 2 & 6 & 3 & 7 \\ 9 & 3 & 8 & 1 \\ 4 & 6 & 9 & 7 \end{bmatrix}$$

$$= \begin{bmatrix} 3+2 & 5+6 & 2+3 & 4+7 \\ 7+9 & 6+3 & 8+8 & 4+1 \\ 5+4 & 1+6 & 3+9 & 8+7 \end{bmatrix}$$

$$= \begin{bmatrix} 5 & 11 & 5 & 11 \\ 16 & 9 & 16 & 5 \\ 9 & 7 & 12 & 15 \end{bmatrix}$$

A.3 MATRIX MULTIPLICATION

Two matrices may be multiplied only if the number of columns in the first matrix is the same as the number of rows in the second matrix. Two matrices will be used to illustrate the multiplication: P and Q.

$$P = \begin{bmatrix} 2 & 6 \\ 7 & 4 \end{bmatrix}$$

$$Q = \begin{bmatrix} 4 & 1 & 3 \\ 6 & 2 & 5 \end{bmatrix}$$

Initially, the matrices are multiplied by taking the first row of the first matrix and multiplying it by the first column of the second matrix. The value corresponding to this calculation is the first element of the first matrix's row multiplied by the first element of the second matrix's column, added to the second element of the first matrix's row multiplied by the second element of the second matrix's column, and so on. In this example, the first row of P (2,6) is multiplied by the first column of Q (4,6), resulting in a value 44 ($2 \times 4 + 6 \times 6$). This calculation is repeated for all combinations of rows from the first matrix by columns of the second matrix, as illustrated in the following example:

$$PQ = \begin{bmatrix} 2 & 6 \\ 7 & 4 \end{bmatrix} \begin{bmatrix} 4 & 1 & 3 \\ 6 & 2 & 5 \end{bmatrix}$$

$$= \begin{bmatrix} 2 \times 4 + 6 \times 6 & 2 \times 1 + 6 \times 2 & 2 \times 3 + 6 \times 5 \\ 7 \times 4 + 4 \times 6 & 7 \times 1 + 4 \times 2 & 7 \times 3 + 4 \times 5 \end{bmatrix}$$

$$= \begin{bmatrix} 44 & 14 & 36 \\ 52 & 15 & 41 \end{bmatrix}$$

A.4 TRANSPOSE OF A MATRIX

For a matrix (A) that corresponds to m rows and n columns, a transpose will exchange the rows and columns (A^T). For example:

$$A = \begin{bmatrix} 7 & 3 \\ 8 & 5 \\ 2 & 9 \end{bmatrix}$$

$$A^T = \begin{bmatrix} 7 & 8 & 2 \\ 3 & 5 & 9 \end{bmatrix}$$

A.5 INVERSE OF A MATRIX

If a matrix A is multiplied by its inverse (A^{-1}), the result will be the identity matrix, where all values are 0, expect the diagonal, which is all 1s. The following illustrates:

$$A^{-1}A = AA^{-1} = \begin{bmatrix} 1 & 0 & \ldots & 0 \\ 0 & 1 & \ldots & \ldots \\ \ldots & \ldots & \ldots & 0 \\ 0 & \ldots & 0 & 1 \end{bmatrix}$$

Calculating the inverse of a square matrix is computationally challenging, and is always done on a computer.

SOFTWARE

B.1 SOFTWARE OVERVIEW

B.1.1 Software Objectives

The Traceis Data Exploration Studio is a software tool for data analysis and data mining. It was developed to provide a hands-on experience demonstrating many of the techniques described in this book, as well as in the preceding book (Myatt, 2007). It incorporates a number of tools for preparing and summarizing data, as well as tools for grouping, finding patterns, and making predictions. These tools are integrated through a series of coordinating views on to the data. Table B.1 summarizes the methods included in the Traceis software.

The software provides tools for:

- *Preparing data*: Having identified a data set to mine, the preparation of the data is often one of the most time-consuming aspects of any data mining project. Understanding what types of data are contained in the data set is important; after obtaining a sense of the data, cleaning, transforming, and appropriately reducing the number of variables and observations helps ensure that the set is ready for any subsequent analysis. Preparing the data helps develop a more thorough understanding of the data. The Traceis software contains tools for loading, cleaning, understanding, and reducing the number of observations and variables, as shown in Table B.1.

- *Tables and graphs*: Visualizing the data can help in preparing the data, to identify patterns and trends, as well as understanding the results of any analysis. Tables and graphs are critical tools for use throughout any data mining project. Table B.1 summarizes the types of data visualization tools included in the software. The software contains a variety of visualization tools. For instance, it includes tools for creating contingency tables that display frequency data. With the software, summary tables that present information on groups of observations are also available. Finally, the software has tools for creating interactive graphs (histograms, frequency polygrams, scatterplots, and box plots) that can be displayed in any order, and it includes tools for combining any number of these possibilities.

- *Statistics*: Statistics play an important role throughout all data mining projects. The software assists in describing the variables in the data set. It also enables

---

Making Sense of Data II. By Glenn J. Myatt and Wayne P. Johnson

TABLE B.1 Summary of Methods Available in the Traceis Software

Method	Tools
Preparing	Loading the data (open), searching the data set (search), characterizing variables (characterize), removing observations and variables (remove), cleaning the data (clean), transforming variables (transform), segmenting the data set (segment), and principal component analysis (PCA)
Tables and graphs	Contingency table, summary table, graphs, and graph matrices
Statistics	Descriptive statistics, confidence intervals, hypothesis tests, chi-square test, ANOVA, and comparative statistics
Grouping	Clustering, association rules, decision trees
Prediction	Linear regression, discriminant analysis, logistic regression, naive Bayes, k-nearest neighbors (kNN), classification and regression trees (CART), and neural networks

statements to be made about the data with confidence. A number of descriptive, inferential, and comparative statistical analyses are provided, and these analyses are summarized in Table B.1.

- *Grouping*: Grouping data is another essential part of many data mining projects. It can assist in preparing the data, as well as in segmenting a data set. Clustering is a flexible approach to grouping observations, and a variety of clustering techniques have been implemented in the Traceis software. Other grouping methods are available in the software, including a technique called associative rules, which groups observations and then identifies "if ... then ..." rules concerning the data. For example, if customers buy product A and product B, then they generally buy product D as well. Another tool is the decision tree. The decision tree method is effective for grouping observations into a hierarchical tree. At each branch of the tree, the observations are divided into subsets based on specific criteria, such as customers who buy product A. This division is guided using a variable of interest, such as the profit associated with a particular customer. In this example, customers are divided into profitable or not profitable groups, based on products they purchased. Table B.1 summarizes the grouping methods contained in the Traceis software.

- *Prediction*: The use of historical information to make future decisions is the focus of prediction. A variety of methods have been implemented, as shown in Table B.1, to build prediction models. These methods work in a variety of situations; for example, some methods only predict continuous variables, while other methods predict binary data.

The software allows exploration of these different methods. The software works with the data sets available on the website (http://www.makingsenseofdata.com) and used in these books. It also works on other, private data sets. The different visualization and analysis methods provide views of the data from different angles, enabling a person to gain insight into alternative and complementary approaches to analyzing data.

Figure B.1 Contents of the folder containing the Traceis software

B.1.2 Access and Installation

The Traceis software can be accessed from the website http://www.making-senseofdata.com/. The software is contained in a zipped file, and once downloaded, it can be unzipped into any folder on a computer. In addition to downloading the zipped file, a license key to use the software can be obtained by sending an email to software@makingsenseofdata.com. An email will be sent to you containing the key, which is simply a number.

To run the software, double click on the Traceis.jar file, as shown in Fig. B.1. The first time the software is run, the user will need to enter the license key number mentioned in the email sent. Also contained in the folder with the software is a subfolder called "Tutorial datasets," which contains sample data sets to use with the software, along with a description of the data sets.

The associated website (http://www.makingsenseofdata.com/) also contains the current minimum requirements for running the software, which can be used on any computer with the Java Virtual Machine (JVM) loaded. Generally, this JVM comes installed on most computers; however, it can be downloaded from Sun's Java website (http://www.java.com/), if necessary. Information on updates to the software will also be provided on the website (http://www.makingsenseofdata.com/).

B.1.3 User Interface Overview

The Traceis user interface is divided into five areas, as shown in Fig. B.2, summarized as:

- *Steps*: Each of the steps in the four-step process outlined in Sections 1.2–1.5 of this book, that is, (1) definition, (2) preparation, (3) analysis, and (4) deployment, is presented on the left of the main user interface window. Any option under each step can be selected. Figure B.2 shows an annotated screenshot from the Traceis software, where the "Grouping" option has been selected from the "Steps" panel.

- *Analyses*: The alternative analysis methods available are organized based on the option selected in the "Steps" panel. As depicted on Fig. B.2, the "Grouping" option was selected, and three alternative analyses are now available, as shown in the tabs "Clustering," "Associative rules," and "Decision tree." In this example, "Clustering" was selected.

- *Tools*: The "Tools" area of the screen shows the parameters and options for implementing any selected analysis of the data. For example, different parameters to run a cluster analysis are shown in Fig. B.2.

- *Results*: The "Results" area of the screen contains the results of any analysis. Figure B.2, for example, displays the result of this cluster analysis.

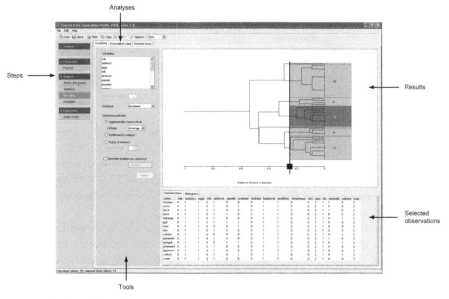

Figure B.2   Overview of the Traceis software

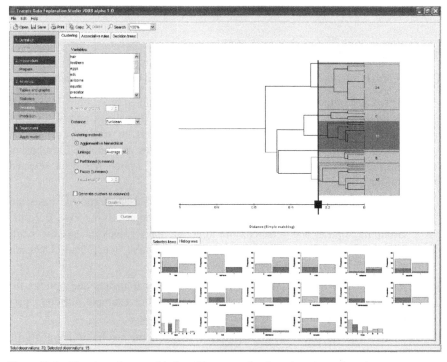

Figure B.3   Traceis software with the "Histograms" selected to show the subset

An interactive hierarchical dendrogram is presented, and, in this example, the number of clusters can be adjusted, and the contents of individual clusters can be selected. Here, 15 observations were selected.

• *Selected observations*: Figure B.2 also displays the selected set of observations on the same screen. There are two options for displaying subsets: "Selected items" and "Histograms." Under "Selected items," a table corresponding to only those items selected is shown. A "Histogram" option can also be selected, such as that shown in Fig. B.3, in which each variable is shown as a separate histogram. The lighter shaded bars represent the frequency distribution for all values, and the darker bars represent just the frequency for the subset selected. The histogram view is also interactive and a single bar or multiple bars can be selected. Single bars are selected by clicking once on a bar. Multiple bars are selected by lassoing those bars of interest. In addition, the total number of selected observations is shown in the bottom left-hand corner of the full user interface, alongside the total number of observations in the data set.

B.2   DATA PREPARATION

B.2.1   Overview

Preparing data for analysis is a critical and time-consuming phase in any data mining exercise. There are a number of tools available in Traceis to streamline and aid the preparation of data (see Fig. B.4), including methods for asking questions of and

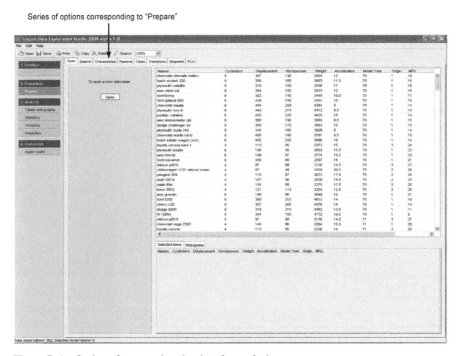

Figure B.4   Options for preparing the data for analysis

understanding the data, such as searching capabilities and principal component analysis. The preparation of the data may include removing observations and variables, as well as creating new columns of data. The following section details the tools available in the Traceis software for preparing data.

B.2.2 Reading in Data

The first step in executing any analysis is to load data into the system. The data should contain the observations, each comprising a series of variables. The software will read from a text file where each observation is on a separate line. As a matter of convenience, many data files include the variables' names as the first line in the file. Each observation should be divided into a series of values corresponding to the variables. A specific separator or delimiter should separate each individual value, such as a comma or a tab. The following provides an example of the content of a text file containing a data table:

Names,Cyclinders,Displacement

chevrolet chevelle malibu,8,307

buick skylark 320,8,350

plymouth satellite,8,318

The first row is the column headings, and each subsequent row is the individual observations. The values are consistently separated using a comma.

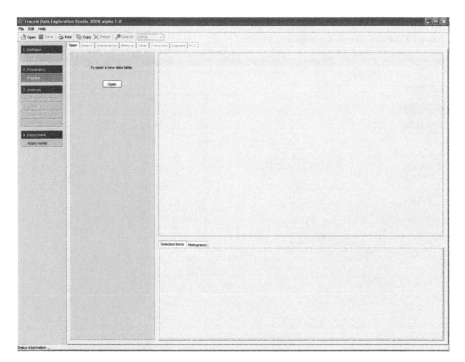

Figure B.5 Opening a data table

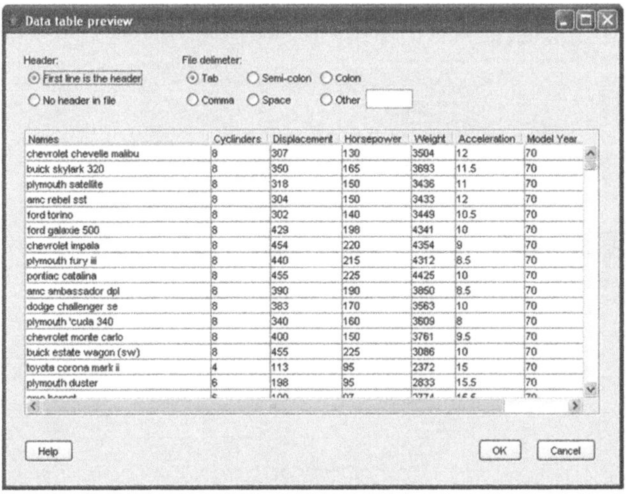

Figure B.6   Preview window to ensure the data is correctly read in

Selecting the "Open" button, as shown in Fig. B.5, starts the process of loading a data set. The first step in loading the data is to review the data to make sure it is formatted into the correct rows and columns. A preview window is initially displayed, as shown in Fig. B.6. The software makes some assumptions about the data, such as assuming that the first line in the table is the column headers (where appropriate), and assuming that certain delimiters (such as commas or tabs) will be used to separate the observation's values. If these assumptions are incorrect, they can be overridden using the options available on the "data table preview" screen, as shown on Fig. B.6. If no header is contained in the text file, the software will automatically assign a header to each column [Variable(1), Variable(2), and so on]. The table can be sorted in ascending order by clicking on any of the column headers, thus enabling easy review of the highest and lowest values in each column, or viewing if the variable contains any text values. Once the data appears to be assigned as rows and columns correctly, clicking on the "OK" button will load the data into the Traceis software.

B.2.3   Searching the Data

Once a data set has been loaded into Traceis, it can be searched in a number of ways by selecting the "Search" tab. The tools, as shown in Fig. B.7, are provided to search one or more terms, where each term is comprised of the variable to search, an operator ($=$, $<$, $>$, and $\neq$), as well as a value for which to search.

   To search for a single term, the "Number of search terms" option should be set to 1. Searches such as "Names $=$ ford torino," or "mpg $<$ 20" can be initiated. When multiple search terms are chosen, a boolean operator can be set to either "AND" or "OR," allowing for the construction of searches such as "mpg $>$ 20 AND mpg $<$ 30," which will return observations where mpg is between 20 and 30. Alternatively, the search terms could include different variables, such as "mpg $>$ 30 OR Weight $>$ 3000,"

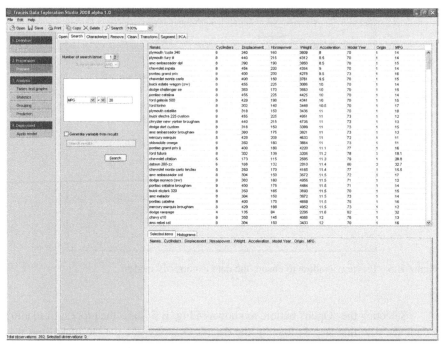

Figure B.7   Searching the variable mpg for observations greater than 20

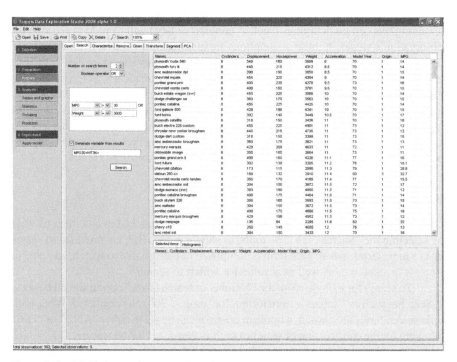

Figure B.8   Multiple search terms selected

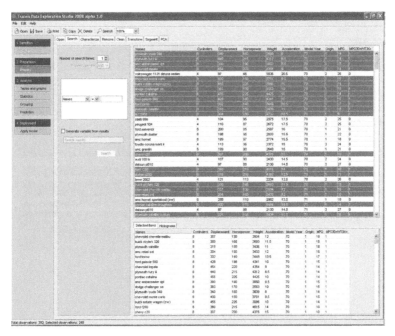

Figure B.9   New variable added to the data table generated from the search results

which will return all observations where mpg is greater than 30 as well as all observations where weight is greater than 3000. These concepts are shown in Fig. B.8.

A new dummy variable can also be generated from the results, where values of 1 correspond to the presence of those observations identified in the search and 0 indicates the absence. The name of this new variable is set by checking the box "Generate variable from results," and entering a name. This new variable is then generated once the search is completed. This dummy variable has a number of uses, including partitioning the data based on subsets generated from specific searches, creating a set of observations to test a hypothesis, and generating new terms for use in a prediction model. In addition, combining more than one of these dummy variables enables more complex queries to be specified, which allows combinations of "AND" and "OR" operations.

Once the search query has been entered, clicking on the "Search" button will start the search. The resulting observations are displayed in the subset area, as well as highlighted in the display area, as shown in Fig. B.9. If the "Generate variable from results" option was selected, a new variable will be added to the data table.

B.2.4   Variable Characterization

When the data is loaded into the Traceis software, the variables are analyzed and assigned to various categories. This assignment can be viewed for each variable, by clicking on the "Characterization" tab. An example of this characterization is shown in Fig. B.10.

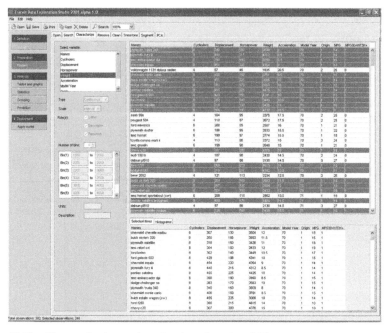

Figure B.10 Characterization options for the selected variable

The categories include type (dichotomous, continuous, and discrete), scale (nominal, ordinal, interval, and ratio) and role (label, independent variable, and response), as discussed in Section 1.3.4. "Labels" refers to a nominal variable where the majority of observations are different, and hence could not be used within a predictive model. However, labels are still useful to describe the individual observations. For continuous variables, each observation is assigned to a range of values or bins, which are displayed in Fig. B.10.

B.2.5 Removing Observations and Variables

The "Remove" tab provides tools to remove observations or variables from the data table. Observations can be selected from most pages throughout the program and may be removed by either clicking on the "Delete" button at the top of the user interface or from the "Remove" tab. Constants or selected variables can also be removed. Figure B.11 illustrates the "Remove" tools, and this screen allows the user to remove selected observations, all constants, or any selected variables. The table in the main display will be modified after the "Remove" button has been clicked in accordance with the selected changes.

B.2.6 Cleaning the Data

For variables containing missing data or nonnumeric data, a series of cleaning options are available from the "Clean" tab. Figure B.12 illustrates some available options.

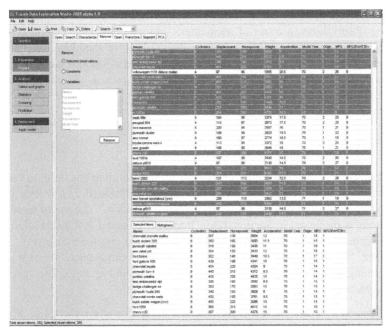

Figure B.11   Tools for removing observations and variables

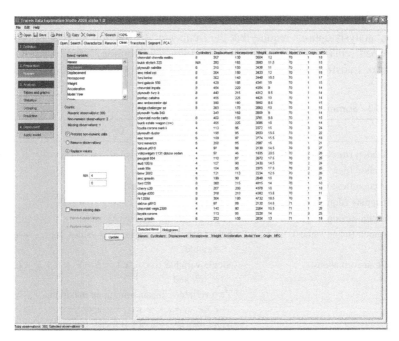

Figure B.12   Options for cleaning a variable

Once a variable is selected, the following summaries are provided: a count of the numeric observations, a count of nonnumeric observations, and a count of observations with missing data. Nonnumeric observations in the data can either be removed or replaced by a numeric value. A similar set of options are available for handling missing data. Once the variable has been updated, the changes will be reflected in the results area of the updated table. Section 1.3.5 discusses these options in more detail.

B.2.7 Transforming the Data

The "Transform" tab offers a number of ways for transforming one or more variables into a new variable. These tools are summarized in Table B.2, and they can be selected from the drop-down "Select type of transformation." The use of these transformations is discussed in Section 1.3.6. It should be noted that all transformation options generate a new variable, and do not replace the original variable(s).

The "Normalization (new range)" option provides three alternatives for transforming a single variable to a new range, as shown in Fig. B.13. The three transformations are:

- *Min−max*: This option will generate a new variable where the values map into a specific range identified in the "From:" and "To:" fields.

- *z-score*: This option generates a new variable where the values are normalized based on the number of standard deviations above or below the mean.

- *Decimal scaling*: This transformation moves the decimal to ensure the range is between -1 and $+1$.

Certain analysis options require that the frequency distribution reflects a normal, or bell-shaped curve. The "Normalization (new distribution)" option provides a number of transformations that generate a new frequency distribution for a variable,

TABLE B.2 Summary of Transformations Available in Traceis

Tool	Options
Normalization (new range)	Min−max, z-score, decimal scaling
Normalization (new distribution)	log, $-$log, box-Cox
Normalization (text-to-numbers)	Values for selected variable
Value mapping (dummy variables)	Variables
Discretization (using ranges)	Select ranges for selected variable
Discretization (using values)	Select new values for existing values, for the selected variable
General transformations	Single variable: $x \times c$, $x + c$, $x \div c$, $c \div x$, $x - c$, $c - x$, x^2, x^3, and $\sqrt{x}$
	Two variables: mean, minimum, maximum, $x + y$, $x + c$, $x \times c$, $x \div y$, $y \div x$, $x - y$, and $y - x$
	More than two variables: mean, minimum, maximum, sum, and product

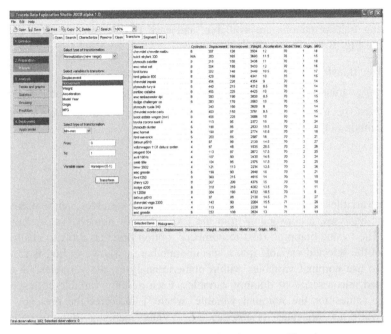

Figure B.13 Window illustrating the "Normalization (new range)" options

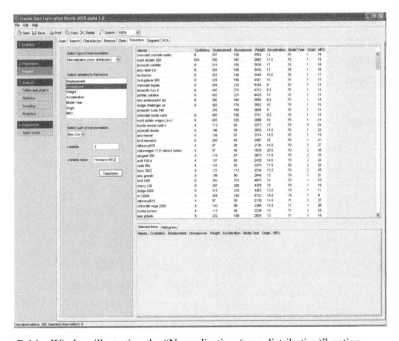

Figure B.14 Window illustrating the "Normalization (new distribution)" option

which can be viewed in the subset window after selecting the "Histograms" tab. The following transformations are available, and each can be selected from the tools shown in Fig. B.14. They include:

- *log*: transforms the data using a log (base 10) transformation;
- *−log*: transforms the data using a negative log (base 10) transform;
- *box-Cox*: transforms the data with the following formula, where a value for lambda (λ) must be specified.

$$value' = \frac{value^\lambda - 1}{\lambda}$$

Certain variables contain text values, and before these variables can be used within numeric analyses, a conversion from the text values to numeric values must take place. The "value mapping (text-to-number)" option provides tools to change each value of the selected variable into a specific number, as shown in Fig. B.15.

To use nominal variables within numeric analyses, the variables are usually converted into a series of dummy variables. Each dummy variable corresponds to specific values for the nominal variable, where 1 indicates the presence of the value while a 0 indicates its absence. The "Value mapping (dummy variables)" tool can be applied to any variable containing text values, and it will automatically generate a series of variables, as shown in Fig. B.16. It should be noted that, when using dummy variables with certain analyses, such as linear regression or logistic

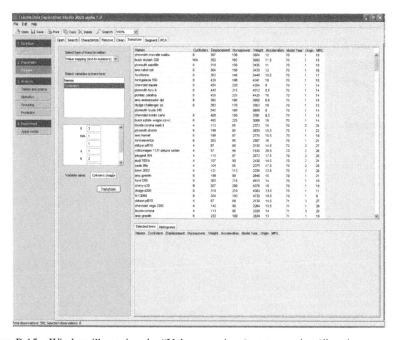

Figure B.15   Window illustrating the "Value mapping (text-to-numbers)" option

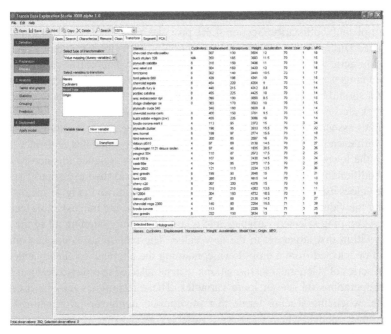

Figure B.16   Window illustrating the "Value mapping (dummy variables)" option

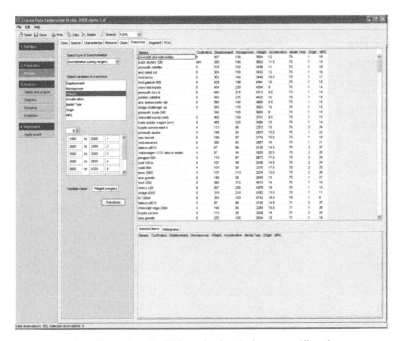

Figure B.17   Window illustrating the "Discretization (using ranges)" option

regression, the use of all generated variables should be avoided because of the problem of multicolinearity. Section 4.3.1 provides more details on this concept.

The "Discretization (using ranges)" option provides tools for converting a continuous numeric variable into a series of discrete values. This conversion is based on specified ranges, as shown in Fig. B.17. Having selected a single variable, the number of ranges should be set, along with the lower and upper bounds for each range. Once the ranges are set, this tool substitutes the old, continuous numeric variable with the new values associated with the ranges in which the values falls. A value is also associated with each range, and after the transformation any value that is greater than or equal to the lower bound or less than the upper bound is assigned this new value.

Additionally, any categorical variable can be transformed to a smaller series of discrete values using "Discretization (using values)," as shown in Fig. B.18. Instead of grouping the values into ranges, as in "Discretiziation (using ranges)," this technique involves grouping selected variables into a larger set, and assigning all of the observations within that larger set to the new value. The individual values can either be typed in or selected from a drop-down containing the alternatives already entered.

A series of "General transformations" can be selected in order to perform mathematical operations on one or more variables. These techniques may be useful, for example, when introducing terms that incorporate nonlinear information for use with a linear model, when a series of variables need to be averaged, or when an aggregated variable needs to be generated. Having selected "General transformation," one or more variables can be selected, as shown in Fig. B.19. To select a single variable, click once on the desired variable name, and to select multiple variables

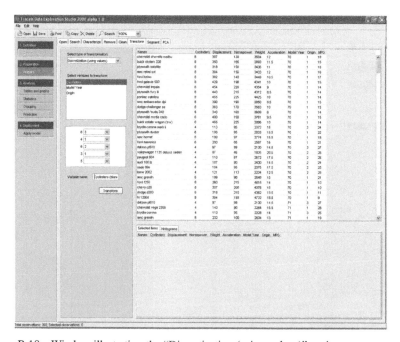

Figure B.18   Window illustrating the "Discretization (using values)" option

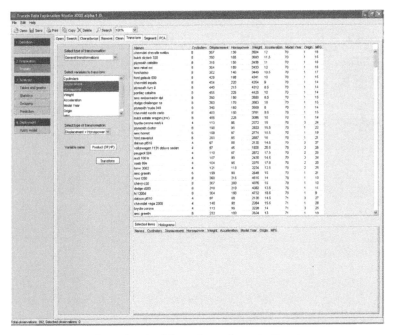

Figure B.19   Window illustrating the "General transformation" option

use the $<$ctrl$>$ + click to select contiguous items and $<$shift$>$ + click to select continuous items.

When a single variable has been selected, the following common mathematic operations are available where x refers to the selected variable and c is a specified constant: $x \times c$, $x + c$, $x \div c$, $c \div x$, $x - c$, $c - x$, x^2, x^3, and $\sqrt{x}$.

When two variables, x and y, have been selected, in addition to the mean, minimum, and maximum functions, the following mathematical transformations are available: $x + y$, $x + c$, $x \times c$, $x \div y$, $y \div x$, $x - y$, and $y - x$.

When more than two variables are selected, the following operations can be applied to the values: mean, minimum, maximum, sum, and product.

By combining the results of any of these transformations, more complex formulas can be generated.

B.2.8   Segmentation

In some situations, generating a smaller subset of observations may be necessary to conduct an efficient analysis of the data, as described in Section 1.3.8. In Fig. B.20, the "Segment" tab has been selected and the number of observations to be included in the subset is set. The software provides two options for selecting a subset of observations:

- *Random*: This option will select the specified number of observations randomly. Each observation in the original set has an equal chance of being included in the new set.

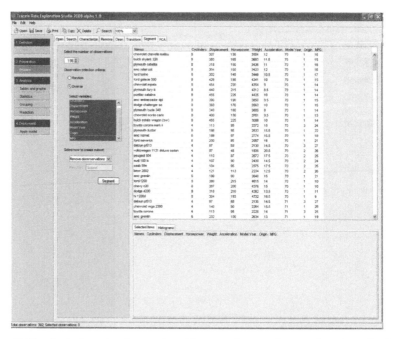

Figure B.20   Analysis option for creating either a random or diverse subset from the data

- *Diverse*: This option will identify the desired number of observations that are representative of the original set. This selection is achieved by initially clustering the observations using k-means clustering, using the euclidean distance where k is the target number of observations in the subset. Once the observations are clustered, a single observation is selected from each cluster that is the closest to the center of the cluster.

At this point only the type of selection process has been selected; how the subset is actually generated must now be set. There are two options selected from the "Select how to create subset" drop-down: (1) generating a new set data set containing only the subset ("Remove observations"); or (2) generating a new dummy variable where 1 represents the inclusion of the observation in the new subset, and 0 indicates its absence ("Generate dummy variable"). If option (2) is selected, the name of this new dummy variable should be provided.

B.2.9   Principal Component Analysis

Principal component analysis (PCA) analyzes the linear relationships between variables and attempts to extract a smaller number of factors that represent the variance in the data, as described in Section 4.2. Figure B.21 displays the software screen after the "PCA" tab is selected, showing the tools available. To select variables to analyze, <ctrl> + click will select contiguous ranges of variables, <shift> + click will select continuous ranges, and <ctrl> + A will select all variables. Once

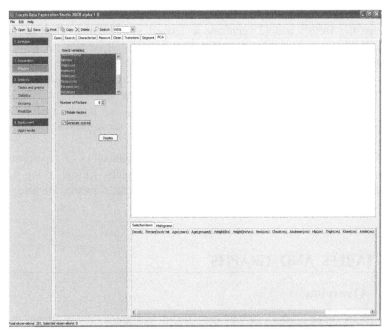

Figure B.21   Tools to generate a principal component analysis

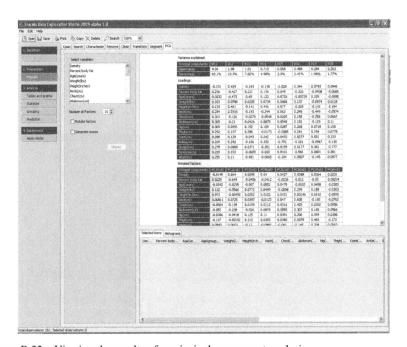

Figure B.22   Viewing the results of a principal component analysis

selected, the number of derived factors should be specified, as well as options to (1) rotate the factors and (2) generate a derived score. If the "Generate scores" option is selected, a series of principal components will be generated, and these components correspond to the number of factors identified. These new variables will have the names "PC-1," "PC-2," and so on for the original principal components. If the "Rotate factors" option is selected, the new variables will be named "PC-1(rot)," "PC-2(rot)," and so on. Clicking the "Display" button will perform a principal component analysis, and the analysis will be shown in the results area, as shown in Fig. B.22.

For each of the principal components, the eigenvalues and percentage of variance attributable to each of the principal components is listed, along with the loadings for each of the principal components. The rotated factors are also shown, if this option was selected.

B.3 TABLES AND GRAPHS

B.3.1 Overview

The tools to create a series of summary tables and graphs are available from the "Tables and Graphs" option, as shown in Fig. B.23. These analyses options are available from the tabs highlighted in the figure, and they are: "Contingency table," "Summary table," "Graphs," and "Graph matrices." Table B.3 provides a summary

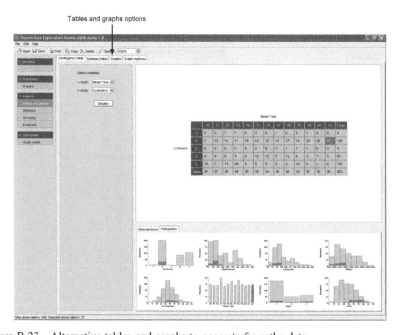

Figure B.23   Alternative tables and graphs to generate from the data

TABLE B.3   Summary of Tables and Graphs

	Variables	Displays
Tables		
Contingency table	2 categorical	Frequency counts
Summary table	1 categorical; any numeric to summarize	Summaries of groups
Graphs		
Histograms	1 variable	Frequency distribution
Scatterplots	2 numeric variables	Relationships
Frequency polygrams	1 variable	Frequency distribution
Box plots	1 continuous variable	Distribution
Graph matrices	Any number of variables	Multivariate analysis

of the graphs and tables available, along with the type of data that can be used and the types of information that can be displayed.

B.3.2   Contingency Tables

A contingency table, as discussed in Section 2.3, can be generated from the "Contingency table" option. The *x*- and *y*-axes of the table are both categorical variables, and they can be selected from the tool options, as shown in Fig. B.24.

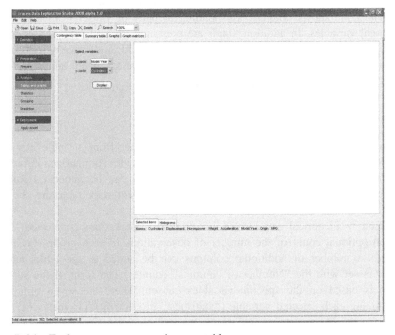

Figure B.24   Tools to construct a contingency table

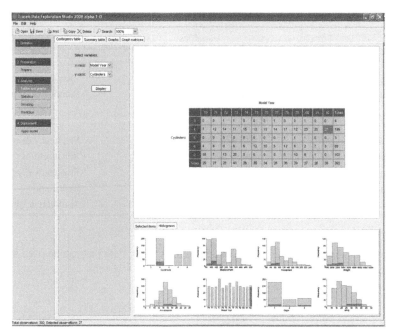

Figure B.25   Viewing and selecting a set of observations from a contingency table

Having selected the *x*- and *y*-axes, clicking on the "Display" button will generate a contingency table in the results window, as shown in Fig. B.25.

The table shows counts corresponding to pairs of values from the selected categorical variables. In addition, totals are presented for each row and column in the table. Each of the cells in the table can be selected, and, once selected, those observations will be displayed in the selected observations panel and other places throughout the software.

B.3.3   Summary Tables

A summary table groups observations using the values from a single categorical variable, as discussed in Section 2.3. In addition, it provides summarized information concerning each of these groups. A summary table can be generated from the "Summary table" tab, and the tools for building a summary table are shown in Fig. B.26.

First, a categorical variable for grouping the observations is selected. At this time, an optional count of the number of observations in each group can also be selected. A number of additional columns can be added to the table, and this number is set with the "Number of columns" counter. The form of the summary created is based on the specific variables chosen. There are seven options for summarizing each selected variable ("Statistics:" drop-down): (1) mean, (2) mode, (3) minimum, (4) maximum, (5) sum, (6) variance, and (7) standard deviation. Clicking on the "Display" button will generate a summary table corresponding to

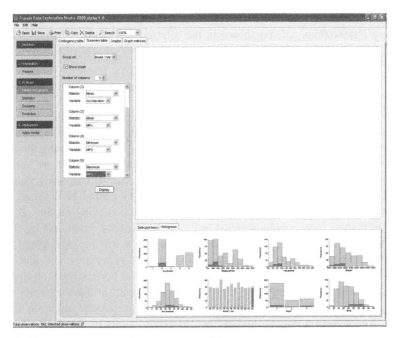

Figure B.26   Tools for generating a summary table

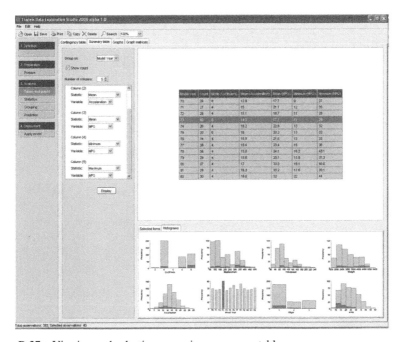

Figure B.27   Viewing and selecting a row in a summary table

the options selected, as shown in Fig. B.27. Individual rows can be selected, and the resulting observations will be shown in the selected observations panel and other places throughout the program.

B.3.4   Graphs

Multiple graphs can be shown on a single screen to summarize the data set, as discussed in Chapter 2. These specific graphs are generated from the "Graphs" tab, and the tools to define these graphs are shown in Fig. B.28. After selecting the desired number of graphs, a series of options for each graph is provided. There are four types of graphs: (1) histogram, (2) scatterplot, (3) box plot, and (4) frequency polygram. In addition, the variable or pair of variables to display should be selected. Once the collection of graphs to display has been determined, clicking on the "Display" button will show these graphs selected in the results area, as shown in Fig. B.29.

In each of the graphs, any selected observations will be highlighted on all graphs with darker shading. The histograms, frequency polygrams, and scatterplots are all interactive, allowing for interaction with the data. For instance, observations can be selected by clicking on a histogram bar or a point in the scatterplot or frequency polygram. In addition, a lasso can be drawn around multiple bars or points. Any selection will be updated on the other graphs in the results areas (as shown in Fig. B.30), as well as being made available in other analysis options throughout the program.

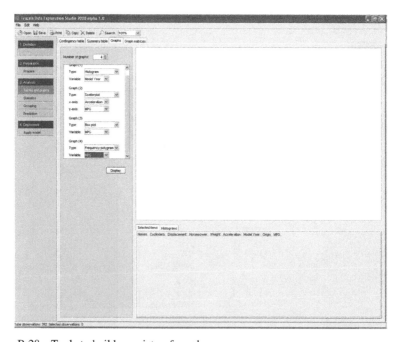

Figure B.28   Tools to build a variety of graphs

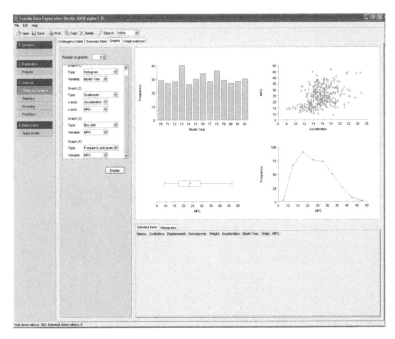

Figure B.29   View of the selected graphs

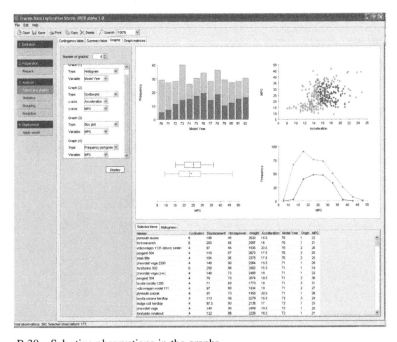

Figure B.30   Selecting observations in the graphs

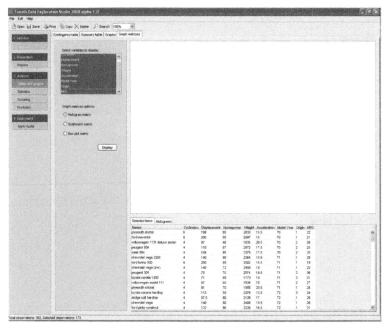

Figure B.31 Tools for generating a histogram, scatterplot, or box plot matrix

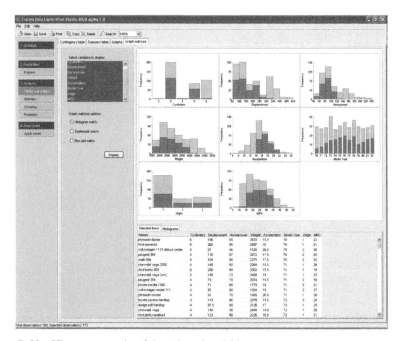

Figure B.32 Histogram matrix of the selected variables

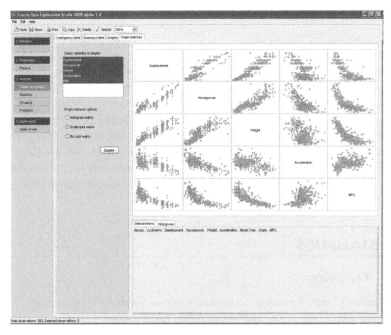

Figure B.33   Scatterplot matrix of all the selected variables

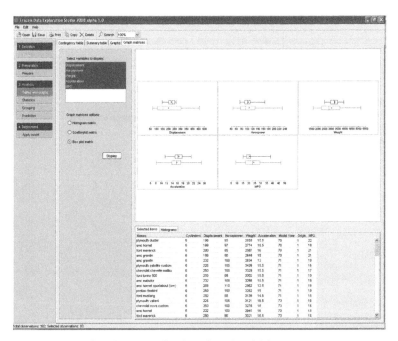

Figure B.34   Matrix of the box plots for the selected variables

B.3.5 Graph Matrices

The "Graph matrices" tab presents a series of specific graphs in a table, as discussed in Chapter 2. The tools for generating this matrix of graphs are shown in Fig. B.31. The tools provide options to display a histogram, scatterplot, or a box plot matrix for the selected variables. Multiple variables can be selected using <ctrl> + click for contiguous variables, <shift> + click for continuous variables and <ctrl> + A for all variables. Clicking on the "Display" button will show the matrix in the results area.

Figures B.32–B.34 show a histogram, scatterplot, and box plot matrix, respectively. The histogram and box plot matrices present the graphs only for the selected variables. In contrast, the scatterplot matrix shows scatterplots for all combinations of the variables selected. The names of the scatterplot axes are shown in the boxes where no graphs are drawn.

B.4 STATISTICS

B.4.1 Overview

The "Statistics" option provides a series of methods for describing variables, making statements about the data, and quantifying relationships between variables, as discussed in Myatt (2007). The software provides a number of options located in the various tabs along the top of the main window: "Descriptive," "Confidence intervals," "Hypothesis tests," "Chi-square," "ANOVA," and "Comparative," as shown in Fig. B.35.

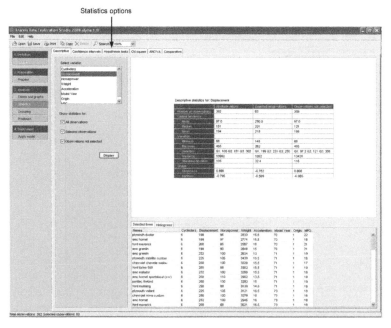

Figure B.35   Statistics options presented as tabs

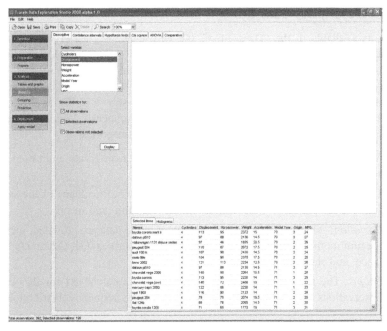

Figure B.36 Tools summarizing a single variable using a series of descriptive statistics

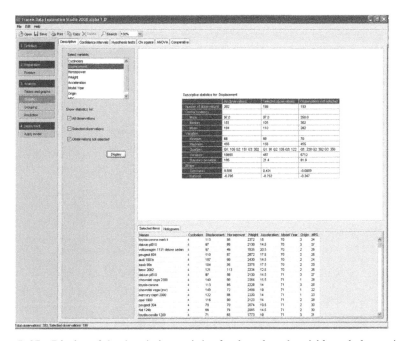

Figure B.37 Display of the descriptive statistics for the selected variable and observations

B.4.2 Descriptive Statistics

The tools available in the "Descriptive" tab will generate a variety of descriptive statistics for a single variable, as shown in Fig. B.36. For the selected variable, descriptive statistics can be generated for (1) all observations, (2) the selected observations, and (3) observations not selected. Clicking on the "Display" button presents the selected descriptive statistics in the results area, as shown in Fig. B.37.

For each of the sets of observations selected, a number of descriptive statistics are calculated. They are organized into the following categories:

- *Number of observations*: The number or count of observations in each of the sets is enumerated.

- *Central tendency*: Measures that quantify the central tendency of the selected variable are provided, including the mode, medium, and mean.

- *Variation*: Measures that quantify variation in the data are provided, including the minimum value, maximum value, the three quartiles (Q1, Q2, and Q3), variance, and standard deviation.

- *Shape*: The skewness and kurtosis estimates are calculated for the selected observations in order to quantify the shape of the frequency distribution.

B.4.3 Confidence Intervals

The "Confidence intervals" analysis calculates an interval estimate for a selected variable that is based on a specific confidence level, as shown in Fig. B.38. In

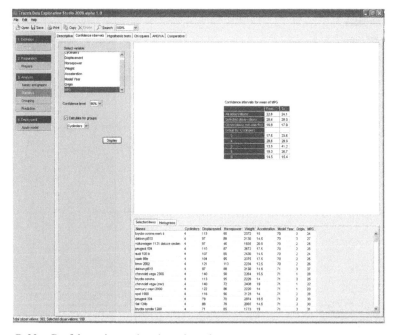

Figure B.38   Confidence interval tools and results

addition, confidence intervals for groups of observations, defined using a single categorical variable, can also be displayed. The confidence intervals for the variables, as well as any selected groups, can be seen in the results area, as shown in Fig. B.38.

B.4.4 Hypothesis Tests

The "Hypothesis tests" analysis performs a hypothesis test on a single variable. When a categorical variable is selected, the hypothesis test takes into consideration the proportion of that categorical variable with a specific value. This value must be set with the "Where x is:" drop-down option, where x is the selected categorical variable. When a continuous variable is selected, the hypothesis test uses the mean. The test can take into consideration one or two groups of observations.

When the "single group" option is selected, as shown in Fig. B.39, the members of this group should be defined. The four options are: (1) all observations in the data table, (2) those selected observations, (3) those observations not selected, and (4) those observations corresponding to a specific value of a categorical variable. The confidence level, or α, should be selected, and possible options are 0.1, 0.05, or 0.01. The hypothesis test should then be described for the selected observations. The value for the null hypothesis should be set, as well as information concerning the alternative hypothesis, that is, whether the alternative should be greater than, less than, or not equal to.

When two groups of observations are selected, as shown in Fig. B.40, both groups should be defined. The three options are: (1) those selected observations, (2) those observations not selected, and (3) those observations corresponding to a

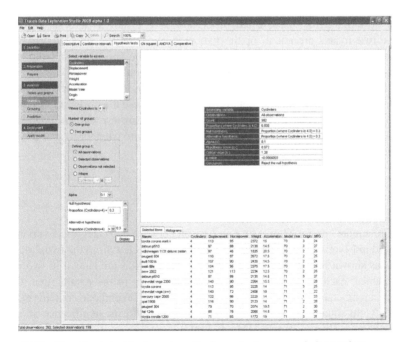

Figure B.39   Hypothesis test tools and results for a single group of observations

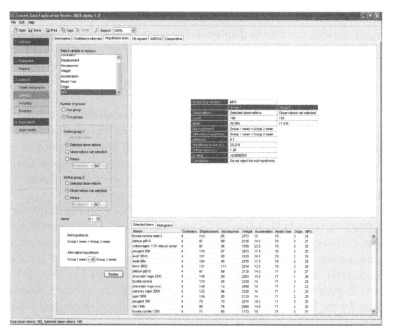

Figure B.40   Hypothesis test tools and results for two groups of observations

specific value of a categorical variable. As before, the confidence level, or α, should be selected from the following list: 0.1, 0.05, or 0.01. The specific hypothesis test should be defined for the selected observations. The null hypothesis is set such that the two means are equal, or the two proportions of the selected variable are the same. Again, the alternative hypothesis should be specified, and the options are: less than, greater than, or not equal to the null hypothesis.

The results of the hypothesis test are presented in the results area, as shown in Fig. B.40. These results include details of the variable and the observations assessed, including the mean value or values, the actual hypothesis test with confidence level, as well as the z-score, the critical z-score, the p-value, and whether to accept or reject the null hypothesis.

B.4.5   Chi-Square Test

The "Chi-square" option allows for comparison between two categorical variables, as discussed in Myatt (2007). These two variables are selected, from the two drop-down menus, as shown in Fig. B.41. The results of the analysis are shown in two contingency tables, each of which is shown simultaneously in the results area. One of the contingency tables includes the actual data, and the second one contains expected results. A chi-square assessment is also calculated, and its significance should be tested against a chi-square table (Myatt, 2007).

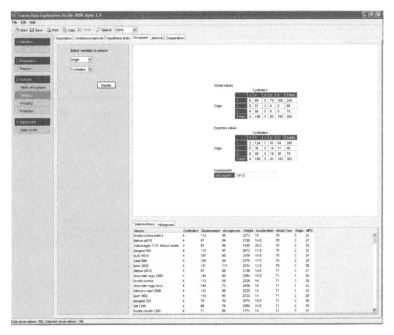

Figure B.41   Chi square tools and results

B.4.6   ANOVA

The "ANOVA" option assesses the relationship for a particular variable between different groups. The ANOVA tools are shown in Fig. B.42. The variable to assess and the categorical variable used to group the observations should be selected. A confidence level, or α, should also be assigned. The resulting analysis is presented in the results area, as shown in Fig. B.42.

The results area shows the groups identified using the selected categorical variable, in addition to the number of observations in each group, the mean value for each group, and the variance for each groups. The mean square within and between is calculated, along with the F-statistic, which should be tested against the critical F-score using an F-table (see Myatt, 2007) based on the following: α, degrees of freedom (within), and degrees of freedom (between).

B.4.7   Comparative Statistics

The linear relationship between a series of continuous variables can be assessed using the "Comparative" statistics option. The tools available for this option are shown in Fig. B.43. For the selected variables, this analysis will calculate a correlation coefficient and, if selected, a squared correlation coefficient for each pair of variables selected. The tables are presented in the results area, as illustrated in Fig. B.43.

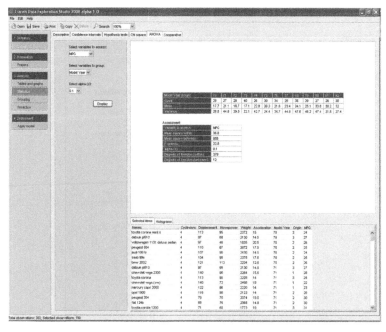

Figure B.42 ANOVA tools and results

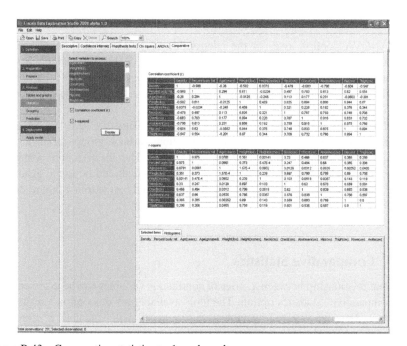

Figure B.43 Comparative statistics tools and results

B.5   GROUPING

B.5.1   Overview

A number of analysis options are available for grouping observations into sets of related observations, as shown in Fig. B.44. These options are: "Clustering," "Associative rules," and "Decision trees," and these options are summarized in Table B.4.

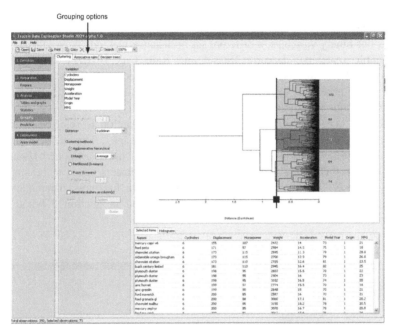

Figure B.44   Grouping options available as tabs

TABLE B.4   Summary of the Grouping Options

Grouping option	Input variables	Response	Options
Agglomerative hierarchical clustering	Any numeric	None	Joining rule Distance measures
k-means clustering	Any continuous	None	Number of clusters Distance measures
Fuzzy k-means clustering	Any continuous	None	Number of clusters Distance measures Fuzziness (q)
Associative rules	Any categorical	None	"THEN" restrictions Value restrictions
Decision trees	Any categorical or continuous	Single numeric or text	Minimum node size

B.5.2 Clustering

A number of methods for clustering observations are available in the Traceis software, including agglomerative hierarchical clustering, partitioned (k-means) clustering, and fuzzy (fuzzy k-means) clustering. These options are selected from the clustering tools menu, as shown in Fig. B.45.

All of these clustering methods require the selection of one or more variables, as well as the selection of a distance measure used to compare observations. For numeric variables (not binary), the following distance calculations are available: euclidean, square euclidean, Manhattan, maximum, Minkowski ($\lambda = 3$), Minkowski ($\lambda = 4$), Minkowski ($\lambda = 5$), Mahalanobis, Canberra, correlation coefficient, and Gower. For binary variables, a different set of distance methods is provided, including: simple matching, Jaccard, Russel and Rao, Dice, Rogers and Tanimoto, and Gower. These distance options are described in Section 3.2. It should be noted that it is not necessary to normalize the data to a standard range, as the software will perform this step automatically.

If the "Agglomerative hierarchical clustering" option is selected, a linkage method must be chosen from the list: average linkage, complete linkage, and single linkage, as described in Section 3.3. For both the "Partitioned (k-means)" and the "Fuzzy (k-means)" options, the number of clusters needs to be specified. Finally, "Fuzzy (k-means)" clustering also requires a fuzziness parameter to be specified. Partitioned k-means clustering is described in Section 3.4 and fuzzy clustering is described in Section 3.5.

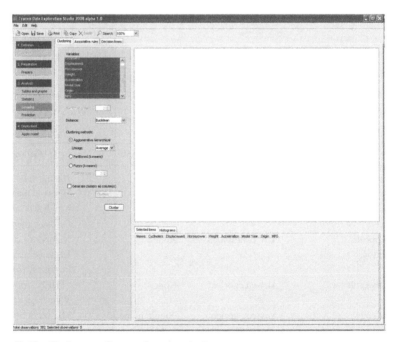

Figure B.45   Tools to perform a cluster analysis

The assignment of the observations to specific clusters can be stored as a separate column or a series of columns in the table by selecting the "Generate clusters as column(s)" option. In the case of agglomerative hierarchical clustering and *k*-means clustering, a single column will be generated, whereas for fuzzy *k*-means clustering, the software will generate one column per cluster containing the membership score for each cluster.

Having specified the type of clustering required, clicking on the "Cluster" button will generate the clusters and the results will be summarized in the results area. Figures B.46–B.48 illustrate agglomerative hierarchical clustering, *k*-means clustering, and fuzzy *k*-means clustering, respectively.

When the agglomerative hierarchical clustering is selected, the results are displayed as a dendrogram describing the hierarchical organization of the data, as shown in Fig. B.46. A vertical line dissects the dendrogram, thus creating clusters of observations to the right of the vertical line. A rectangle is placed around each cluster, and space permitting, a number indicating the size of the cluster is annotated on the right. When the data set has a label variable, clusters with only a single observation are replaced by the label's value (space permitting). The cutoff is interactive; it can be moved by clicking on the square towards the bottom of the cutoff line and moving it to the left or right. Moving the line changes the distance at which the clusters are generated, and hence the number of clusters will change as the cutoff line moves. If the "Generate clusters as column(s)" option is selected, the column in the data table describing the cluster membership will also change.

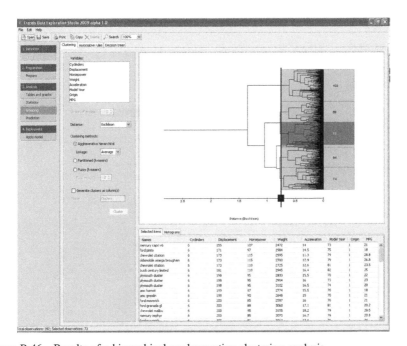

Figure B.46 Results of a hierarchical agglomerative clustering analysis

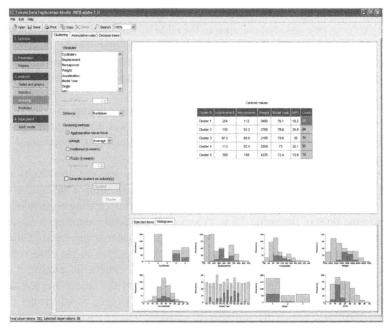

Figure B.47   Results of a *k*-means cluster analysis

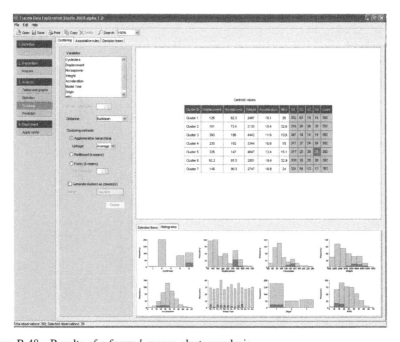

Figure B.48   Results of a fuzzy *k*-means cluster analysis

If the "Partitioned (*k*-means)" clustering option is selected, the results are presented in a table, where each row represents a single cluster, as shown in Fig. B.47. The centroid values for each cluster are presented next to the number of observations in the cluster. This "number of observations" can be selected, and those selected observations are displayed in the selected observations panel, as well as throughout the program.

When "Fuzzy *k*-means" clustering is selected, the results are shown in a table where each row corresponds to a single cluster, as shown in Fig. B.48. The cluster centroid is displayed to summarize the contents of the cluster. With fuzzy clustering, all observations belong to every cluster, and hence the final observation count column corresponds to the total number of observations in the data table. Four additional count columns are provided where Q1 includes observations with membership scores between 0 and 0.25, Q2 includes observations with membership scores between 0.25 and 0.5, Q3 includes observations with membership scores between 0.5 and 0.75, and Q4 includes observations with membership scores greater than 0.75. Counts in these four columns can also be selected, and these selected observations are shown in the selected observations panel and throughout the program.

B.5.3 Associative Rules

The "Associative rules" option will group observations into overlapping groups that are characterized by "If . . . then . . . " rules. This method is described in Myatt (2007). The tools for generating the associative rules are shown in Fig. B.49. A set of

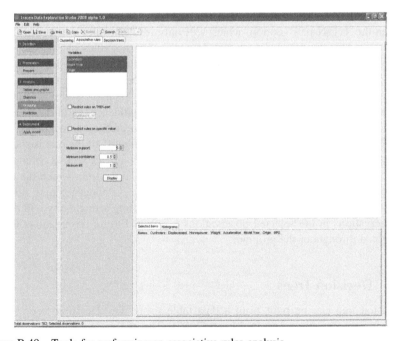

Figure B.49   Tools for performing an associative rules analysis

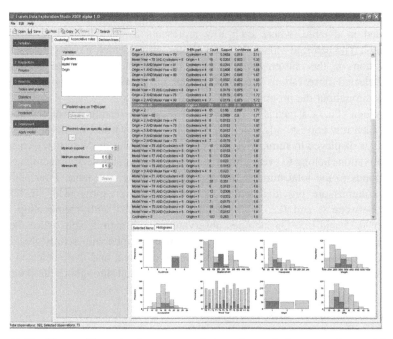

Figure B.50   The resulting rules from an associative rules analysis presented as a table

categorical variables can be selected, and specific values corresponding to these variables will be used in the generated rules. The software includes a "Restrict rules on the THEN-part" option, which will only result in rules where the THEN-part incorporates the selected variable. Also, the "Restrict rules by specific value" function allows for the selection of an appropriate value from the drop-down list. This option is particularly useful when the rules are being generated from a series of dummy variables, and only rules with values of "1" contain useful information. Generating rules with minimum values for support, confidence, and lift can also be set. The resulting rules are shown as a table in the results area, as shown in Fig. B.50.

The results are displayed in a table, where the "IF-part" of the rule is shown in the first column, and the "THEN-part" of the rule is shown in the second column. The next column displays a count of the number of observations from which the rule is derived. The table also displays values for support, confidence, and lift. The table can then be sorted by any of these columns. Selecting a single row will display the observations in the selected observations panel, and those observations will be highlighted throughout the program.

B.5.4   Decision Trees

A decision tree can be built from a data table using the "Decision tree" option. Decision trees are discussed in Myatt (2007). The tools for building a tree are shown in Fig. B.51. Any number of variables can be used as independent variables

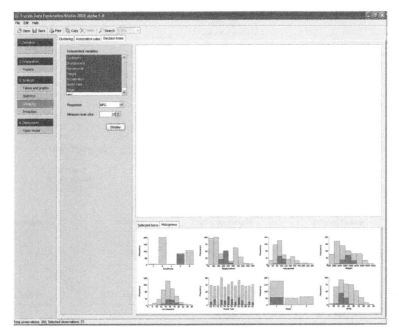

Figure B.51   Tools for performing a decision tree analysis

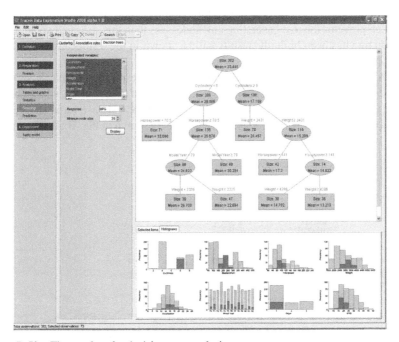

Figure B.52   The results of a decision tree analysis

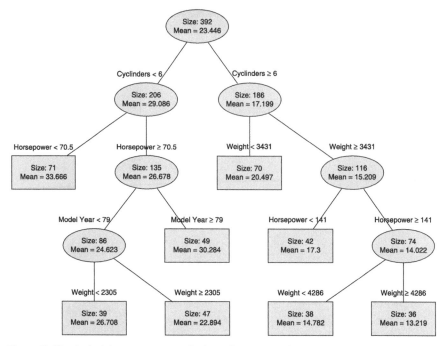

Figure B.53   A decision tree generated where the response is continuous

and a single variable should be assigned as the response. These variables will guide the generation of the decision tree. In addition, a minimum tree node size should be set which prevents the tree generating any nodes less than this specified value. Once a decision tree has been built, the results will be shown in the results area, as shown in Fig. B.52.

A decision tree generated using a continuous response variable is shown in Fig. B.53. The nodes on the decision tree represent sets of observations, which are summarized with a count, along with the average value for the response variable. The result of a decision tree where the response is categorical is shown in Fig. B.54. The number

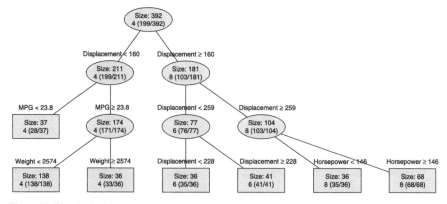

Figure B.54   A decision tree generated where the response is categorical

of observations in the set is shown, along with the most common value, qualified by the number of occurrences of that value compared to the entire node size. In both trees, the criteria used to split the trees are indicated just above the node. Oval nodes represent nonterminal nodes, whereas rectangular nodes represent terminal nodes. The decision trees are interactive; each node can be selected, and the selected observations can be seen below the tree as well as throughout the program.

B.6 PREDICTION

B.6.1 Overview

A series of prediction options are also provided in the Traceis software. As shown in Fig. B.55, the following tools are available: (1) linear regression, (2) discriminant analysis, (3) logistic regression, (4) naive Bayes, (5) kNN, (6) CART, and (7) neural networks. A table summarizing the different prediction options is shown in Table B.5. The ability to do a cross-validation is common to all prediction methods. The results of the cross-validation analysis are presented in a consistent format. Figure B.56 illustrates a cross-validation result for a regression model where the response variable is continuous. The software provides a series of metrics that are calculated and presented in a table, including mean square error, mean absolute error, relative square error, relative absolute error, correlation coefficient, and square correlation coefficient (as described in Section 4.1.3). A scatterplot showing the relationship between the actual response and the predicted values for the cross-validated results is also provided.

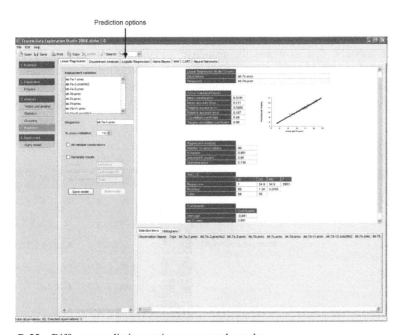

Figure B.55 Different prediction options presented as tabs

TABLE B.5   Summary of the Prediction Options

Method	Independent variables	Response	Options
Linear regression	Numeric	Continuous	
Discriminant analysis	Numeric	Categorical	
Logistic regression	Numeric	Binary	Threshold
Naive Bayes	Categorical	Binary	
kNN	Numeric	Numeric	k
CART	Any	Any	Node size
Neural networks	Numeric	Numeric	Number of cycles Hidden layers Learning rate

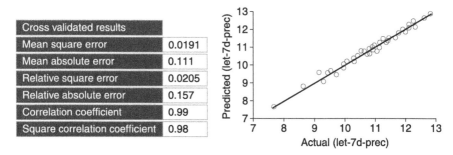

Cross validated results	
Mean square error	0.0191
Mean absolute error	0.111
Relative square error	0.0205
Relative absolute error	0.157
Correlation coefficient	0.99
Square correlation coefficient	0.98

Figure B.56   Example of a regression cross-validated results

Figures B.57 and B.58 provide examples of cross-validation results for classification models, that is where the response variable is categorical. Figure B.57 illustrates the results for a nonbinary classification model. These results include accuracy and error rate metrics and a contingency table of actual response values against predicted values for the cross-validated predictions. Figure B.58 presents an example

Cross validated results	
Accuracy	0.837
Error	0.163

Actual (cyclinders)

Predicted (cyclinders)	3	4	5	6	8	Totals
3	2	0	0	0	0	2
4	0	180	1	22	0	203
5	0	1	0	0	0	1
6	0	0	0	53	10	63
8	2	18	2	8	93	123
Totals	4	199	3	83	103	392

Figure B.57   Example of categorical (nonbinary) cross-validated results

Cross validated results		Actual (high mpg)			
Accuracy	0.908				
Error	0.0918		**1**	**0**	**Totals**
Sensitivity	0.922				
Specificity	0.888	**1**	214	18	232
False positive rate	0.112				
Positive predictive value	0.922	**0**	18	142	160
Negative predictive value	0.888				
False discovery rate	0.0776	**Totals**	232	160	392

(Predicted (high mpg))

Figure B.58 Example of cross-validated results where the response is binary

of a binary classification model where the response can take values 0 or 1. In this example, the following metrics are calculated: accuracy, error rate, sensitivity, specificity, false positive rate, positive predictive value, negative predictive value, and false discovery rate (as described in Section 4.1.5). A contingency table is also provided showing the actual values compared to the cross-validated predicted values.

Another common feature to all the prediction options is the "All variable combinations" option. When this option is selected, all combinations of the selected variables will be used to build a number of models from which the combination that has the highest predictive accuracy is automatically selected as the final model.

B.6.2   Linear Regression

Tools for building multiple linear regression models are found under the "Linear regression" tab. These tools will generate a linear regression model (as discussed

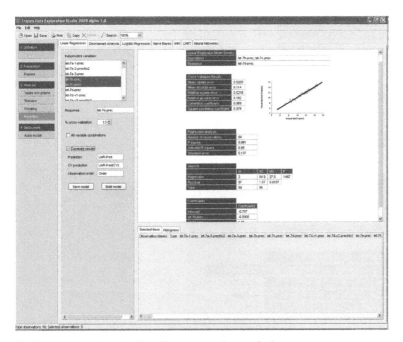

Figure B.59 Tools and results for a linear regression analysis

in Section 4.3) and are, illustrated in Fig. B.59. Any number of independent variables can be selected using $<$ctrl$>$ + click to select contiguous, and $<$shift$>$ + click to select a continuous ranges of variables. A single continuous response variable should be also selected. The cross-validation percentage should be set to indicate the amount of data to set aside for testing. To further analyze the results, a series of new variables can be generated: "Prediction," "CV Prediction," and "Order." A final multiple linear regression model will be built from the entire data set, and the "Prediction" variable will have a prediction for all observations from this model. "CV Prediction" represents the cross-validated prediction, where the predicted values are calculated using a model built from other observations. "Order" contains a number for each observation reflecting the order the observation appears in the data set.

Once a model is built, the results are displayed in the results area, as shown in Fig. B.59. The independent variables and the response variable are initially summarized, along with a summary of the cross-validated results. The software provides a regression analysis summarizing the model accuracy, including R^2, adjusted R^2, and standard error. An ANOVA analysis is generated showing the degrees of freedom (d.f.) of the regression and the residual, along with the mean square (MS), the sum of

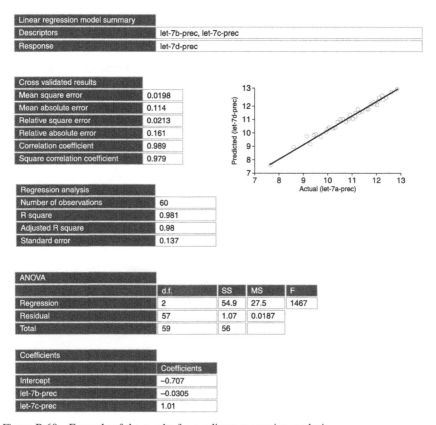

Figure B.60   Example of the results from a linear regression analysis

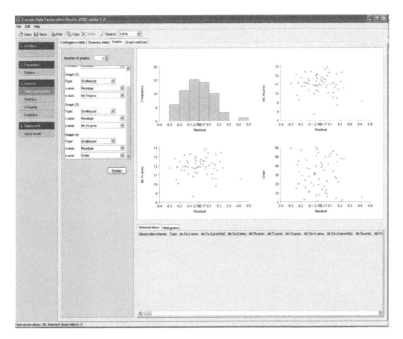

Figure B.61   Analysis of the residuals

squares (SS), and the f-statistic, which should be compared with an f-table to derive a p-value. Finally, the coefficients of the equation are presented. This information is presented in Fig. B.60.

To evaluate the model in more detail, a residual variable can be generated using the "Transform" tab option under "Preparation." To create a residual variable, first select a "General Transformation" and select the actual response and the prediction, and then select "Actual−Prediction." This data can be plotted in the "Graphs" tab to analyze the model further, as shown in Fig. B.61.

Once a model is built and evaluated, it can be saved permanently. The model is saved by clicking on the "Save model" button, as shown in Fig. B.59. A file name should be provided and the model will be saved to a file for future use with other data sets.

B.6.3   Discriminant Analysis

Discriminant analysis models (described in Section 4.4) can be built from the tools available under the "Discriminant analysis" tab. Any numeric independent variables can be selected, as well as any single categorical response variable, as shown in Fig. B.62. The cross-validation percentage should be specified to indicate the data necessary to be set aside when building the multiple cross-validated models. Two variables can also be generated: (1) the prediction from the final model and (2) the cross-validated prediction.

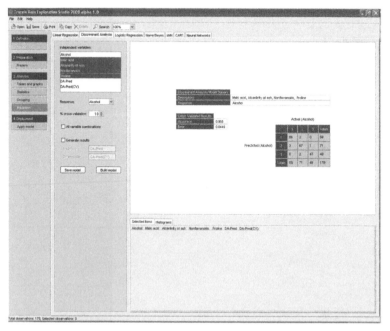

Figure B.62   Tools and results from a discriminant analysis

A summary of the model built is shown in Fig. B.62. This summary displays the independent variables and the response variable, as well as metrics from the cross-validated assessment, described earlier.

Any model generated can be saved to use later with other data sets, using the "Save model" button shown in Fig. B.62.

B.6.4   Logistic Regression

The "Logistic regression" option tab enables the generation of a logistic regression model (described in Section 4.5). Models can only be built for binary response variables; however, any numeric variable can be used as an independent variable. Figure B.63 shows an example of the tools and results of a logistic regression analysis.

A logistic regression model generates a probability, and the prediction is generated from this probability using a specified threshold value. Observations above this threshold value will be assigned a prediction of 1, and those below will be assigned a prediction of 0.

In addition to the cross-validated percentage to set aside, a number of predicted values can be optionally generated by selecting the "Generate results" option: "Prediction," "Prediction prob," "CV Prediction," and "CV Prediction prob." They are the predicted value along with the probability the response is 1 for the final model ("Prediction" and "Prediction prob") and the cross-validated models ("CV Prediction" and "CV Prediction prob").

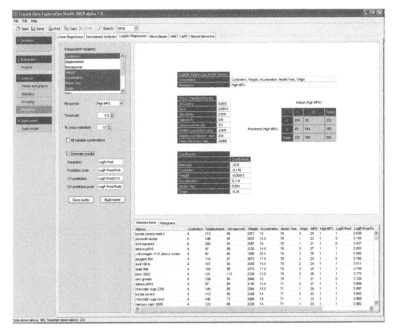

Figure B.63   Tools and results from a logistic regression analysis

Once a model has been built, the results are displayed in the results area, as shown in Fig. B.63. The independent variables and the response variable are initially summarized, along with a summary of the cross-validated results. Finally, the coefficients of the logistic regression equation are also presented.

Any model generated can be saved for use with other data sets, using the "Save model" button shown in Fig. B.63.

B.6.5   Naive Bayes

The "Naive Bayes" tab allows for the generation of a naive Bayes model (see Section 4.6 for more details). Under this implementation, models can only be built for binary response variables with any categorical variable used as an independent variable. These should be set as shown in Fig. B.64.

In addition to the cross-validated percentage to set aside, a number of predicted values can be optionally generated by selecting the "Generate results" option: "Prediction," "Prediction prob," "CV Prediction," and "CV Prediction prob," as discussed in Section B.6.4.

Once a model is built, the results are displayed in the results area, as shown in Fig. B.64. The independent variables and the response variable are initially summarized, along with a summary of the cross-validated results. Any model generated can be saved for use with other data sets, using the "Save model" button, as shown in Fig. B.64.

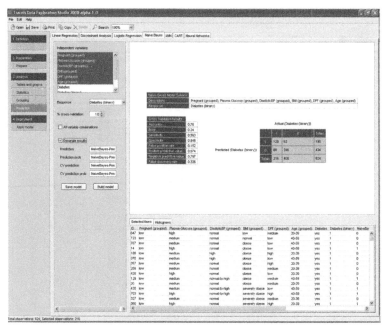

Figure B.64   Tools and results from a naive Bayes analysis

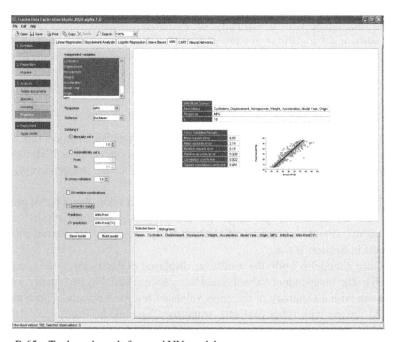

Figure B.65   Tools and result from a *k*NN model

B.6.6   *k*NN

The "*k*NN" analysis tab lets you build *k*-nearest neighbor models (see Myatt, 2007 for more details). Models can be built for any response variable, and any numeric variables can be selected as independent variables, as shown in Fig. B.65. It is not necessary to normalize the data to a standard range as the software will do this automatically. The distance metric should be selected from the drop-down menu. A value for *k* can be set manually. Alternatively, a range can be selected such that the Traceis software builds all models between the lower and upper bounds and selects the model with the smallest error.

In addition to the cross-validated percentage to set aside, a number of predicted values can be optionally generated by selecting the "Generate results" option: "Prediction" and "CV Prediction," which are the predicted values for the final model, along with the predicted values from the cross-validated models.

Once a model has been built, the results are displayed in the results area, as shown in Fig. B.65. The independent variables and the response variable are initially summarized, as well the value for the *k*-nearest neighbor parameter that was either set manually or automatically derived. A summary of the standard cross-validated results is presented. Any model generated can be saved for future use with other data sets, using the "Save model" button shown in Fig. B.65.

B.6.7   CART

The "CART" analysis tab is used to build models based on either a regression tree or a classification tree (see Myatt, 2007 for more details). Models can be built for any response variable and any independent variables, and these variables should be set,

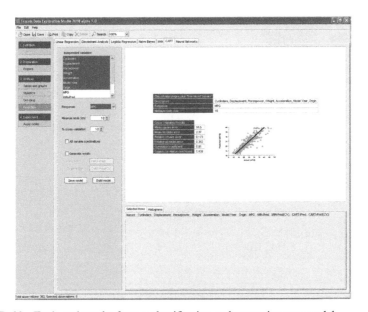

Figure B.66   Tools and results form a classification and regression tree model

as shown in Fig. B.66. Decision trees are generated for the models, and the "Minimum node size" should be set which will prune the tree generation process based on this threshold.

In addition to the cross-validated percentage to set aside, a number of predicted values can be optionally generated by selecting the "Generate results" option: "Prediction" and "CV Prediction," as discussed in Section B.6.6.

Once a model has been built, the results are displayed in the results area, as shown in Fig. B.66. The independent variables along with the response variable are initially summarized, as well as the value used for minimum node size. A summary of the standard cross-validated results is presented. Any model generated can be saved for later use with other data sets, using the "Save model" button from Fig. B.66.

B.6.8   Neural Networks

The "Neural networks" analysis tab allows for the generation of a neural network model. These models can be built for any numeric response variables, and any numeric variable can be set as independent variables, as shown in Fig. B.67. A number of parameters need to be set, including the number of learning cycles the network should perform (sometimes referred to as epochs), the number of hidden layers in the network (between the input layer and the output layer), as well as the learning rate. It is not necessary to normalize the data to a standard range as the software automatically does this. A more detailed description of these parameters is given in Myatt (2007).

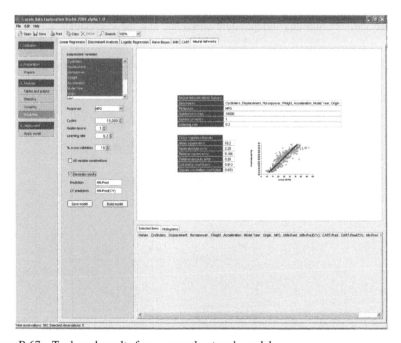

Figure B.67   Tools and results from a neural network model

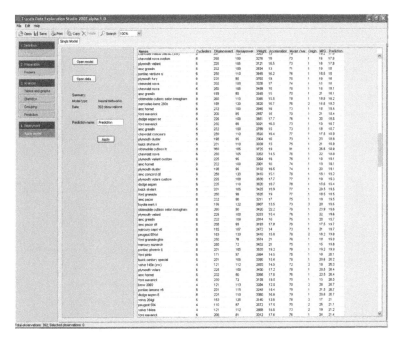

Figure B.68   Model applied to a data set

In addition to the cross-validated percentage to set aside, a number of predicted values can be optionally generated by selecting the "Generate results" option: "Prediction" and "CV prediction," which are the predicted values for the final model, along with the predicted values from the cross-validated models.

Once a model is built, the results are displayed in the results area, as shown in Fig. B.67. The independent variables and the response variable are initially summarized, as well as the values for the model parameters: number of cycles, number of hidden layers, and learning rate. A summary of the standard cross-validated results is presented. Any model generated can be saved for use with other data sets, using the "Save model" button from Fig. B.67.

B.6.9   Apply Model

Any model built and saved can be used with a new data set by selecting the "Apply model" option under the "4. Deployment" step. The model and the new data set should be opened, and a summary of the model and the data is shown. Selecting the "Apply" button will generate a prediction for the observations. The column headings must match those used to build the model. Figure B.68 shows a model applied to a data set.

BIBLIOGRAPHY

Agresti A (2002). *Categorical Data Analysis*. Hoboken, NJ: John Wiley & Sons.

Aldenderfer MS, Blashfield RK (1984). *Cluster Analysis*. Newbury Park, CA: Sage Publications.

Alhaji R, Gao H, Li X, Li J, Zaiane OR, editors (2007). *Advanced Data Mining and Applications. Third International Conference. ADMA 2007*, Harbin, China, August 6–8, 2007, Proceedings. Berlin: Springer.

Allison PD (1998). *Multiple Regression: A Primer*. Thousand Oaks, CA: Pine Forge Press.

Andrews NO, Fox EA (2007). *Recent Developments in Document Clustering*. Blacksburg, VA: Department of Computer Science, Virginia Tech. Available at http://eprints.cs.vt.edu/archive/00001000/01/docclust.pdf. Accessed May 19, 2008.

Balakrishnan N. (1992). *Handbook of the Logistic Distribution*. New York: Marcel Dekker.

Baxevanis AD, Francis Ouellette BF, editors (2005). *Bioinformatics: A Practical Guide to the Analysis of Genes and Proteins*, 3rd edn. Hoboken, NJ: John Wiley & Sons.

Becker RA, Cleveland WS, Shyu M (1996). The visual design and control of trellis display. *Journal of Computational and Graphical Statistics* 5(2): 123–155.

Berner ES, editor (2006). *Clinical Decision Support Systems: Theory and Practice*, 2nd edn. New York: Springer.

Berry MJA, Lindoff GS (2004). *Data Mining Techniques: For Marketing, Sales, and Customer Relationship Management*, 2nd edn. Hoboken, NJ: John Wiley and Sons.

Bertin J (1983). *Semiology of Graphics: Diagrams, Networks, Maps*. Madison, WJ: The University of Wisconsin Press.

Blower PE, Verducci JS, Lin S, Zhou J, Chung JH, Dai Z, Liu CG, Reinhold W, Lorenzi PL, Kaldjian EP, Croce CM, Weinstein JN, Sadee W (2007). MicroRNA expression profiles for the NCI-60 cancer cell panel, *Mol. Cancer Ther.* 6(5): 1483–1491.

Braha D, editor (2001). *Data Mining for Design and Manufacturing: Methods and Applications*. Dordrecht: Kluwer Academic.

CellMiner website (2008). Available at http://discover.nci.nih.gov/cellminer/. Accessed May 19, 2008.

Chen H, Fuller SS, Friedman C, Hersh W, editors (2005). *Medical Informatics: Knowledge Management and Data Mining in Biomedicine*. New York: Springer Science & Business Media.

Cleveland WS (1993). *Visualizing Data*. Summit, NJ: Hobart Press.

Cleveland WS (1994). *The Elements of Graphing Data*. Summit, NJ: Hobart Press.

Cohen JE (2004). Mathematics is biology's next microscope, only better; biology is mathematics' next physics, only better. *PLoS Biology* 2(12): 2017–2023. Available at www.plosbiology.org. Accessed May 19, 2008.

Making Sense of Data II. By Glenn J. Myatt and Wayne P. Johnson
Copyright © 2009 John Wiley & Sons, Inc.

Cristianini N, Shawe-Taylor J (2000). *An Introduction to Support Vector Machines and Other Kernel-Based Learning Methods*. Cambridge: Cambridge University Press.

Dalby A, Nourse JG, Gushurst AK, Hounshell WD, Grier L, Leland BA, Laufer J (1992). Description of several chemical structure file formats used by computer programs developed at Molecular Design Limited. *Journal of Chemical Information and Computer Sciences* 32: 244–255.

Dasu T, Johnson T (2003). *Exploratory Data Mining and Data Cleaning*. Hoboken, NJ: John Wiley and Sons.

Davenport TH, Harris JG (2007). *Competing on Analytics: The New Science of Winning*. Boston, MA: Harvard Business School Publishing.

Draper NR, Smith H (1998). *Applied Regression Analysis*. Hoboken, NJ: John Wiley and Sons.

Everitt BS, Landau S, Leese M (2001). *Cluster Analysis*. Oxford: Oxford University Press.

Fan W, Wallace L, Rich S, Zhang Z (2006). Tapping the power of text mining. *Communications of the ACM* 49(9): 76–82.

Feldman R, James S (2007). *The Text Mining Handbook: Advanced Approaches in Analyzing Unstructured Data*. New York: Cambridge University Press.

Fielding AH (2007). *Cluster and Classification Techniques for the Biosciences*. New York: Cambridge University Press.

Forsyth R (1990). Zoo database. Available at: http://archive.ics.uci.edu/ml/machine-learning-databases/zoo/zoo.names. Accessed May 19, 2008.

Fox J (1997). *Applied Regression Analysis, Linear Models and Related Methods*. Thousand Oaks, CA: Sage Publications.

Freedman D, Pisani R, Purves R (1998). *Statistics*, 3rd edn. New York: Norton.

Gan CM, Wu J (2007). *Data Clustering: Theory, Algorithms, and Applications*, 1st edn. Philadelphia, PA: Society for Industrial and Applied Mathematics.

Gasteiger J, Engel T, editors (2003). *Chemoinformatics: A Textbook*. Weinheim: Wiley-VCH.

Giudici P (2003). *Applied Data Mining: Statistical Methods for Business and Industry*. Chichester: John Wiley & Sons.

Grossman RL, Kamath C, Kegelmeyer P, Kumar V, Namburu R. (2001). *Data Mining for Scientific and Engineering Applications*. Dordrecht: Kluwer Academic Publishers.

Han J, Kamber M (2005). *Data Mining: Concepts and Techniques*, 2nd edn. San Francisco, CA: Morgan Kaufmann.

Hand DJ, Yu K (2001). Idiot's Bayes—not so stupid after all? *International Statistical Review* 69(3): 385–399.

Hassoun MH (1995). *Fundamentals of Artificial Neural Networks*. Cambridge, MA: The MIT Press.

Hastie T, Tibshirani R, Friedman JH (2003). *The Elements of Statistical Learning*, 3rd edn. New York: Springer Science & Business Media.

Haykin S (1998). *Neural Networks: A Comprehensive Foundation*, 2nd edn. Upper Saddle River, NJ: Prentice Hall.

Hearst MA (1999). Untangling text data mining. In: *Proceedings of the 37th Annual Meeting of the Association for Computational Linguistics on Computational Linguistics (College Park, Maryland, June 20–26, 1999). Annual Meeting of the ACL*. Morristown, NJ: Association for Computational Linguistics.

Hoaglin DC, Mosteller F, Tukey JW (2000). *Understanding Robust and Exploratory Data Analysis*. Hoboken, NJ: John Wiley and Sons.

Hornick MF, Marcade E, Venkayala S (2006). *Java Data Mining: Strategy, Standard, and Practice: A Practical Guide for Architecture, Design, and Implementation*. San Francisco, CA: Morgan Kaufmann.

Hosmer DW, Lemeshow S (2000). *Applied Logistic Regression*, 2nd edn. Hoboken, NJ: John Wiley and Sons.

Huberty CJ (1994). *Applied Discriminant Analysis*. Hoboken, NJ: John Wiley and Sons.

Jackson JA (1991). *A User's Guide to Principal Components*. Hoboken, NJ: John Wiley and Sons.

Jajuga K, Sokolowski A, Bock HH, editors (2002). *Classification, Clustering and Data Analysis*, 1st edn. Berlin: Springer.

Jarvis RA, Patrick EA (1973). Clustering using a similarity measure based on shared nearest neighbours. *IEEE Transactions Computing* C-22(11): 1025–1034.

Jobson JD (1992). *Applied Multivariate Data Analysis: Volume II: Categorical and Multivariate Methods*. New York: Springer.

Johnson RA, Wishern DW (1998). *Applied Multivariate Statistical Analysis Tools*, 4th edn. Upper Saddle River, NJ: Prentice Hall.

Joliffe IT (2002). *Principal Component Analysis*, 2nd edn. New York: Springer.

Jones NC, Pevzner PA (2004). *An Introduction to Bioinformatics Algorithms*. Cambridge, MA: The MIT Press.

Kachigan SK (1991). *Multivariate Statistical Analysis*, 2nd edn. New York: Radius Press.

Kaiser HF (1958). The varimax criterion for analytic rotation in factor analysis. *Psychometrika* 23: 187–200.

Kantardzic M, Zurada J, editors (2005). *Next Generation of Data-Mining Applications*. Hoboken, NJ: John Wiley and Sons.

Kaufman L, Rousseeuw PJ (2005). *Finding Groups in Data: An Introduction to Cluster Analysis*. Hoboken, NJ: John Wiley and Sons.

Kelley LA, Gardner SP, Sutcliffe MJ (1996). An automated approach for clustering an ensemble of NMR-derived protein structures not conformationally-related subfamilies. *Protein Engineering* 9(2): 1063–1065.

Klösgen W, Żytkow JM, Zyt J, editors (2002). *Handbook of Data Mining and Knowledge Discovery*. Oxford: Oxford University Press.

Kohonen T (1990). *Self-Organization and Associative Memory*, 3rd edn. New York: Springer.

Kohonen T (2001). *Self-Organizing Maps*, 3rd edn. Berlin: Springer.

Kroeze JH, Matthee MC, Bothma TJ (2003). Differentiating data- and text-mining terminology. In: Eloff J, Engelbrecht A, Kotzé P, Eloff M, editors. *Proceedings of the 2003 Annual Research Conference of the South African Institute of Computer Scientists and Information Technologists on Enablement Through Technology* (September 17–19, 2003), pp. 93–101. South Africa: South African Institute for Computer Scientists and Information Technologists.

Kyrgyzov IO, Kyrgyzov OO, Maître H, Campedel M (2007). Kernel MDL to Determine the Number of Clusters. In: *Machine Learning and Data Mining in Pattern Recognition: Proceedings 5th International Conference*, MLDM 2007, July 18–20, Leipzig, pp. 203–217. Berlin: Springer.

Lachenbruch PA (1975). *Discriminant Analysis*. New York: Hafner Press.

Leach AR, Gillet VJ (2003). *An Introduction to Chemoinformatics*. The Netherlands: Springer.

Li X, Wang S, Dong ZY, editors (2005). *Advanced Data Mining and Applications: First International Conference, ADMA 2005*, Wuhan, July 22–24, 2005, Proceedings. Berlin: Springer.

Li X, Zaiane OR, Li Z, editors (2006). *Advanced Data Mining and Applications: Second International Conference, ADMA 2006*, Xi'an, August 14–16, 2006, Proceedings. Berlin: Springer.

Maimon O, Rokach L, editors (2005). *The Data Mining and Knowledge Discovery Handbook.* Berlin: Springer.

McCue C (2007). *Data Mining and Predictive Analysis: Intelligence Gathering and Crime Analysis.* Oxford: Butterworth-Heinemann.

McLachlan GJ (2004). *Discriminant Analysis and Statistical Pattern Recognition.* New York: Wiley-Interscience.

McNeils PD (2004). *Neural Networks in Finance: Gaining Predictive Edge in the Market.* Burlington, MA: Elsevier Academic Press.

Mean J (2003). *Investigative Data Mining for Security and Criminal Detection.* Burlington, MA: Butterworth Heinemann.

Milligan GW, Cooper MC (1985). An examination of procedures for determining the number of clusters in a data set. *Pschyometrika* 50: 159–179.

Mirkin B (2005). *Clustering for Data Mining: A Data Recovery Approach.* Boca Raton, FL: Chapman & Hall/CRC.

Murphy PM, Aha DW (1994). *UCI Repository of Machine Learning Databases.* Irvine, CA: University of California, Department of Information and Computer Science. Available at http://www.ics.uci.edu/~mlearn/MLRepository.html. Accessed May 19, 2008.

Myatt GJ (2007). *Making Sense of Data: A Practical Guide to Exploratory Data Analysis and Data Mining.* Hoboken, NJ: John Wiley and Sons.

Nargundkar S (2008). Available at http://www.nargund.com/gsu/index.htm. Accessed May 19, 2008.

Netz A, Chaudhuri S, Fayyad U, Bernhardt J (2001). Integrating Data Mining with SQL Databases: OLE DB for Data Mining. In: *17th International Conference on Data Engineering: April 2–6, 2001, Heidelberg, Germany.* Washington, DC: IEEE Computer Society Press.

Pardalos PM, Boginski VL, Vazacopoulos A, editors (2007). *Data Mining in Biomedicine.* Berlin: Springer.

Perner P, editor (2006). *Advances in Data Mining: Applications in Medicine, Web Mining, Marketing, Image and Signal Mining.* Berlin: Springer.

Pyle D (1999). *Data Preparation for Data Mining.* San Francisco, CA: Morgan Kaufmann.

Pyle D (2003). *Business Modeling and Data Mining.* San Francisco, CA: Morgan Kaufmann.

Refaat M (2007). *Data Preparation for Data Mining Using SAS.* San Francisco, CA: Morgan Kaufmann.

Rencher AC (2002). *Methods of Multivariate Analysis*, 2nd edn. Hoboken, NJ: John Wiley and Sons.

Rud OP (2001). *Data Mining Cookbook: Modeling Data for Marketing, Risk and Customer Relationship Management.* Hoboken, NJ: John Wiley and Sons.

Salvador S, Chan P (2004). Determining the number of clusters/segments in hierarchical clustering/segmentation algorithms. In: *Proceedings of 16th IEEE International Conference on Tools with AI.* Washington, DC: IEEE Computer Society.

Scherf U, Ross DT, Waltham M, Smith LH, Lee JK, Tanabe L, Kohn KW, Reinhold WC, Myers TG, Andrews DT, Scudiero DA, Eisen MB, Sausville EA, Pommier Y, Botstein D, Brown PO, Weinstein JN (2000). A gene expression database for the molecular pharmacology of cancer. *Nature Genetics* 4: 236–244.

Schilopf B, Smola AJ (2001). *Learning with Kernels: Support Vector Machines, Regularization, Optimization, and Beyond.* Cambridge, MA: The MIT Press.

Shadbolt J, Taylor JG, editors (2002). *Neural Networks and the Financial Markets: Predicting, Combining and Portfolio Optimisation.* London: Springer.

Shortliffe EH, Cimino JJ, editors (2006). *Biomedical Informatics: Computer Applications in Health Care and Biomedicine.* New York: Springer Science & Business Media.

Shumueli G, Patel NR, Bruce PC (2007). *Data Mining for Business Intelligence: Concepts, Techniques, and Applications in Microsoft Office Excel with XLMiner.* Hoboken, NJ: John Wiley and Sons.

Spence R (2001). *Information Visualization.* New York: ACM Press.

Strang G (2006). *Linear Algebra and its Applications,* 3rd edn. London: Thomson.

Sumathi S, Sivanandam SN (2006). *Introduction to Data Mining and its Applications.* Berlin: Springer.

Tan PN, Steinbach M, Kumar V (2005). *Introduction to Data Mining.* Reading, MA: Addison Wesley.

Thuraisingham B (2003). *Web Data Mining and Applications in Business Intelligence and Counter-Terrorism.* Boca Raton, FL: CRC Press.

Tidwell J (2006). *Designing Interfaces: Patterns for Effective Interaction Design.* Sebastopol, CA: O'Reilly Media.

Tufte ER (1983). *The Visual Display of Quantitative Information.* Cheshire, CT: Graphics Press.

Tufte ER (1990). *Envisioning Information.* Cheshire, CT: Graphics Press.

Victor B (2006). *Magic Ink: Information Software and the Graphical Interface.* Available at http://worrydream.com/MagicInk. Accessed May 19, 2008.

Wainer H (1997). *Visual Revelations: Graphical Tales of Fate and Deception from Napolean Bonaparte to Ross Perot.* New York: Copernicus.

Wainer H (2005). *Graphic Discovery: A Trout in the Milk and Other Visual Adventures.* Princeton, NJ: Princeton University Press.

Weinstein JN, Myers TG, O'Connor PM, Friend SH, Fornace Jr AJ, Kohn KW, Fojo T, Bates SE, Rubinstein LV, Anderson NL, Buolamwini JK, van Osdol WW, Monks AP, Scudiero DA, Sausville EA, Zaharevitz DW, Bunow B, Viswanadhan VN, Johnson GS, Wittes RE, Paull KD (1997). An information-intensive approach to the molecular pharmacology of cancer. *Science* 275: 343–349.

Wilkinson L (2005). *The Grammar of Graphics.* New York: Springer.

Witten IH, Frank E (2005). *Data Mining: Practical Machine Learning Tools and Techniques.* San Francisco, CA: Morgan Kaufmann Publishers.

Xiong J (2006). *Essential Bioinformatics.* New York: Cambridge University Press.

Ypma TJ (1995). Historical development of the Newton–Raphson method. *SIAM Review* 37(4): 531–551.

INDEX

Making Sense of Data II. By Glenn J. Myatt and Wayne P. Johnson
Copyright © 2009 John Wiley & Sons, Inc.

Lightning Source UK Ltd.
Milton Keynes UK
UKOW06f0831100616

276020UK00003B/27/P